Sheikh Nur al Jerrahi
Lex Hixon

ATOM FROM THE SUN OF KNOWLEDGE

LEX HIXON
NUR AL JERRAHI

PIR PUBLICATIONS WESTPORT, CT

Published by Pir Publications, Inc.
256 Post Road East • Westport, Connecticut 06880
(203) 221-7595 • Fax (203) 454-5873

First Edition
ISBN: 1-879708-05-1
Library of Congress Catalog Card Number: 91-067719

Contents

Preface

Friday Noon Prayer is the central focus of the Islamic week, when Prophet Muhammad, upon him be peace, spiritually addresses his global community through the mature *imams,* the knowledgeable leaders of prayer. My beloved Shaykh, Muzaffer Ashqi al-Jerrahi of Istanbul, used to introduce his talks on such solemn occasions with these powerful words: "I am presenting here simply an atom from the sun of knowledge, a drop from the ocean of knowledge."

I once dreamed of my noble Shaykh, dressed regally in robes and turban, speaking from a beautiful wooden *minbar,* the pulpit of the Messenger of Allah located in every holy mosque on the planet. Four Quranic chanters, who were his dervishes, stood around the bottom step of the pulpit. Esoterically interpreted, these were the Prophet Muhammad and his four rightly guided successors: Abu Bakr, Umar, Uthman, and Ali. This dream demonstrates how the original community of Islam, now apparently separated from us by fourteen centuries of linear time, is mystically replicated during each successive generation by authentic shaykhs and their loyal dervishes.

I am western born, liberally educated, with sound background in Greek, European, and Indian philosophy as well as in several religious traditions. In the Islamic year 1400, which was 1980 of the Common Era, I became one of the formal successors of Muzaffer Effendi. I knelt before him, side by side with my spiritual sister Fariha al-Jerrahi, at the Mosque of Divine Ease, the Masjid al-Farah, in New York City. After placing his magnificent green and gold turban upon my head, the Grand Shaykh opened his palms and offered this supplication: "May whatever has come into me from Allah and from the Prophet of Allah now enter into him."

After this brief prayer, Shaykh Muzaffer removed his turban from my head and placed it on the western woman beside me. I would have enjoyed wearing it longer, but spiritual transmission, like turning on an electric light, is instantaneous.

Atom from the Sun of Knowledge is a verbal expression of the ineffable light that flowed into my being during that moment. This mysterious illumination has been raining down within me ever since from the green turban of Nureddin Jerrahi, the Light of Universal Religion, who lived three hundred years ago in Istanbul. My Shaykh, Muzaffer Effendi, was nineteenth in his line of successors.

These writings have manifested through inspiration, combined with literary effort, during the eleven years since this transmission took place. Many of them were composed during the holy month of Ramadan—fasting from sunrise to sunset, feasting and praying until one hour before the first light. This collection represents only a fraction of the spontaneous teachings that the author has presented orally during the eleven years of his responsibility as a Sufi guide, who is a friend to souls and an interpreter of dreams. Combined with its companion volume, *Heart of the Koran,* published in 1988, this work presents a comprehensive mystical interpretation of Islam.

These contemporary writings of a Western initiate are deeply rooted in the authentic traditions of the ancient Dervish Orders of the East, where for many centuries both Muslim men and Muslim women with social and familial responsibilities have been attaining the highest realization—mystic union with Supreme Reality.

These compositions can be tasted like nectar by the soul and tested like gold in the fire of the heart's longing and sincerity. Such teachings belong only to limitless Truth. They are not confined within any limited context of understanding, including that of the author.

The mystical writings collected here are not personal but represent the universal gift of an unbroken spiritual transmission covering fourteen centuries, beginning with the Messenger of Allah in the desert of Arabia. This lineage does not, strictly speaking, originate with the Prophet Muhammad, upon him be peace, since he considered himself a humble inheritor of the vast spiritual wealth of Adam, Noah, Abraham, Isaac, Ishmael, Jacob, Joseph, Solomon, David, Moses, and Jesus, upon them all be peace.

In another sense, however, this prophetic wealth is the lineage of *nur muhammad,* the Muhammadan Light, parallel to the Christian teaching of the Logos. In the Gospels, the beloved Jesus proclaims, "Before Abraham was, I am." In the Hadith, or Oral Tradition of the Prophet, the beloved Muhammad announces from the same ecstatic level of conscious oneness, "I was a Prophet when Adam was still between water and clay."

These contemporary writings flow through the blessings of the sublime Ali, may Allah eternally enlighten his countenance, and his wife, Fatima the Illumined, the magnificent daughter of the Prophet, and their sons, Imam Hassan and Imam Hussain, as well as the entire line of twelve noble descendants of Muhammad.

These writings are infused with the mysterious blessings of the four central poles of Sufism—Sayyid Ahmad Rufai, Sayyid Ahmad Badawi, Sayyid Sultan Abdul Qadir Gaylani, and Sayyid Ibrahim Dusuqi. These writings are fundamentally indebted to the great woman of Islam who opened wide the path of lover and Beloved, the noble Rabia al-Adawiya, may her secret be sanctified. The atmosphere of these writings is permeated, as is all mystical thought and experience in Islam, by the spiritual presence of the King of Lovers, Mansur al-Hallaj, the King of Gnostics, Bayazid Bistami, and the Master of Sobriety, Junayd of Baghdad, may their astonishing spiritual secrets remain well guarded.

Also present through these writings are the sublime gnostic saints Muinuddin Chishti and Shah Naqshiband and their formal successors. Into these two noble lineages as well, the present author has received initiation. These writings fly on the two wings of Sufism, Muhyiddin Ibn Arabi and Mevlana Jelaluddin Rumi, may their spirits be sanctified.

Finally, every movement of heart and mind expressed in these pages was kindled by the Cupbearer of Divine Love for all humanity, Sultan Muhammad Nureddin Jerrahi, the founding Pir of my Order. Nureddin Jerrahi carefully carried on the initiatory line of the Khalwatis, established in Anatolia and settled in Egypt some seven hundred years ago. He gathered together the global riches of Sufism and was recognized by his contemporaries, through spiritual dreams, as the Axis and Seal of Love.

Whatever purity or sincerity is expressed through these writings is the gift of Allah through the intercession of our Pir's noble mother, Amina Taslima. That the book exists at all is due to the blessings of Allah flowing through the prayers of the modern successor to Nureddin Jerrahi, my Shaykh Muzaffer Ashqi, who came to America in 1978 and infused us with a new sense of love and responsibility. He brought with him to New York City the blue sheepskin of Pir Nureddin, which had never left the dervish lodge in Istanbul.

I accompanied Muzaffer Effendi on his eleventh and final pilgrimage to Mecca and Medina in 1980. Yet during the last seven years of his life, this great lover traveled fourteen times to New York City, having fallen in love with the open heart of America. Evidently, the Grand Shaykh found here in the West a greater spiritual priority for his life than in the holy cities of Arabia and the East.

How can I describe eleven years as spiritual guide to various dervish communities in the United States and Mexico? This communion of hearts and minds generated the writings collected here. Each selection the author composed for a certain community, read aloud to that community, discussed and elucidated in the sacramental presence of that community, and revised in the light of such discussion.

So many of my personal and cultural assumptions, and those of my friends in the Jerrahi Order, both eastern and western by birth, have been discarded or transformed. We have made, and must continue to make, subtle adjustments between what is appropriate for a traditional Islamic Order in the ancient East and what is appropriate for a mystical association of liberal-minded persons in the modern West—Muslim by birth, Muslim by adoption, Jewish, Christian, Bahai, Sikh, Hindu, Buddhist, Native American, and nontraditional. We continue to operate, however, with the blessing, guidance, and permission of the present Grand Shaykh of the Jerrahi Order, Sefer Effendi of Istanbul. The history of our community formation will make a significant study sometime in the future.

The maturing process in a Dervish Order is communal. The mystical ascension into Paradise consciousness, and beyond into the Garden of Essence, occurs hand in hand, hearts intertwined eternally. Meeting on Thursday nights for *dhikr,* the dynamic circle of Divine Remembrance, we encounter each other primarily as aspiring souls and only secondarily as personalities with psychological and sociological profiles. The living spiritual documents in this book belong to the landscape of the soul. They cannot remain fixed within any personal, cultural, historical, or religious framework. They are the very energy of essential Reality.

My only prayer is that the writings gathered here will provide an opportunity for further elevation to the sincere hearts of readers, from whatever traditional or nontraditional perspective they may come. Whether *Atom from the Sun of Knowledge* can also contribute to greater planetary appreciation and understanding of the noble way of Islam rests in Divine Foreknowledge alone.

LEX HIXON
NUR AL-JERRAHI
MASJID AL-FARAH
1992/1413

Contents and Purposes

The Introduction to this book, *Four Steps and Seven Levels,* establishes that the author is not operating as a Western intellectual but as a shaykh in the lineage of an ancient Dervish Order, speaking authentically from that living ground. His teaching function is initiatory transmission rather than scholarship. *Atom from the Sun of Knowledge* is not a speculative reinvention of tradition but embodies the continuity of Islamic faith and practice. Islam in all its richness is expressed here through a contemporary mind educated in the precincts of liberal secular humanism, as Moses was raised in the palace of Pharaoh.

The book is divided into two sections: *Traditional Islamic Resources* and *Sufi Inspirations.* This arrangement does not imply any division between the noble tradition of Islam, fourteen centuries of spiritual discipline and exploration, and the rare mystical treasures of the Dervish Orders that are sometimes referred to generically as Sufism. Sacred tradition is an uninterrupted continuum. Meister Eckhart, among the consummate Christian mystics of Europe, whose strong teaching of omniconscious unicity or identity with the Divine is recognizably Sufi, offered his most radical teachings during Sunday sermons to his German congregation. He would have been puzzled or even dismayed by the modern question "Are you Christian or mystic?" Similarly inappropriate is the question I hear again and again: "Are you Muslim or Sufi?" The most refined contemplative experience, the most profound teaching and effective guidance, emerge from the depth and remain within the embrace of carefully transmitted tradition.

Part One, *Traditional Islamic Resources,* concentrates on such subjects as the Daily Prayers, the Holy Quran, the Oral Tradition of the Prophet, the sacred month of Ramadan. One billion Muslims on the planet today would experience agreement of heart with this presentation, regardless of subtle or obvious differences in religious and cultural emphasis.

Part Two, *Sufi Inspirations,* brings the clear principles of Part One into more radical expression, concentrating on distinctive Sufi themes—the path of love, the dervish circle of remembrance, the shaykh or mystic guide, the science of spiritual alchemy, the experience of Paradise during earthly life and entering the Garden of Essence beyond Paradise. These remain deeply Islamic concerns, explored as well by all noble wisdom traditions throughout history. Some contemporary Muslims may experience reservations about the teachings in Part Two, although they will recognize the rootedness of these teachings in the traditional soil of Part One. The grandparents and great grandparents of modern Muslims would have been able to attune more easily to the atmosphere of this book. There has been much cultural and spiritual erosion, and consequent conservative entrenchment, during the last hundred years of Islamic history.

My mode of composition differs dramatically between these two parts of *Atom from the Sun of Knowledge.* Part One was written in English by an author thoroughly schooled and widely read in that language. Part Two was composed by the same author in Spanish for his community of dervishes in Mexico City, writing as a neophyte in this beautiful Western Islamic language, perfumed by many centuries of high Muslim culture in Spain. The author assembled these poems and essays in the manner of collage, working from long lists of rich Spanish words and expressions, allowing them to come together spontaneously. This process of free association, or divinely guided association, produced a form of linguistic expression that would not have been imaginable in English. Incoherencies and errors were eliminated through the editorial assistance of Danielle Garcia Gay, who patiently instructed the author in the grammar, nuance, and rhythm of the Castilian idiom while editing the poems and essays composed in Spanish through this spontaneous process.

Later, the author translated his own writings into English from *Gathering Honey (Recolección de la Miel,* Mexico City, 1989). The results, although somewhat strange to the ear, are unexpectedly revealing. This is why these poems and essays are entitled *Inspirations.* The writings in Part One were also composed in the inspiring ambiance of a mystical Order, often during the sacred month of Ramadan, but they do not display the same intuitive reach as those composed in Spanish. The underlying fervor and creativity of Mexican culture, flowing through both Christian and Pre-hispanic wisdom traditions, nourished the roots of these universal writings. The inherent musicality and fragrance of the Castilian tongue shaped not only the form but the content of these intoxicating songs and manifestos.

The English translations in Part Two benefited from the editorial assistance of Pamela White, who helped clarify and simply the unconventional, torrential language of the Spanish originals. Ms. White has examined with her careful editorial eye the entire manuscript of *Atom from the Sun of Knowledge,* suggesting many subtle changes that have brought the writing to higher clarity and precision. She designed the book's attractive graphic form as well. I also wish to thank Sixtina Friedrich. Her scholarship in Arabic and Turkish combined with her sensitivity to Sufism has been an invaluable resource in insuring accuracy throughout the book.

The purpose of bringing together in one volume both traditional Islamic resources and contemporary Sufi inspirations is to demonstrate the essentially mystical nature of Islam—what may be called its *mystical countenance.* The term *mystical* is used here to indicate the indescribable intimacy between humanity and Divinity suggested by the Quranic phrases "near, nearer than near, and even nearer than that" and "Allah is nearer to us than our central life vein." We are the People of Nearness. Every sensitive Muslim is a dervish. Every unveiled human heart is mystical.

It is my hope that Part One and Part Two will be seen as fully interrelated and experienced as one essential taste. This demonstration of Nearness will enable those born outside Islamic tradition to appreciate its astonishing depth. Perhaps this mere atom from the Sun of Knowledge will help those born and educated inside historical Islam to recover and transmit to future generations the profound spiritual sophistication that belongs to their own global community, which has given birth to more authentic mystics in more diverse societies than any other world religion.

Finally, this book intends to demonstrate that European and post-European culture is a fertile field for the spirit of universality, which is inherent in Islam and which is always propagated by enlightened shaykhs. Humanity is not entering a soulless technological wasteland, as the English poet T. S. Eliot once warned. Nor is any atheistic, authoritarian system of government prevailing on the earth, as once seemed almost inevitable, as least to Marxists. Planetary civilization is evolving instead into the experience of cosmic sacredness and unity so beautifully expressed in the Holy Quran, where Allah calls humanity "the sensitive caretaker of the earthly sphere." I attempt to unfold this Quranic vision more completely in *Heart of the Koran* (Theosophical Publishing House, 1988), which presents meditations on 991 verses from the radiant Book of Reality.

Allah Most High did not design Islam to suppress or supplant other revealed traditions but to safeguard and elucidate their central teaching of omniconscious unicity. The truly human experience of Oneness and its existential implication, compassionate service, is universal Islam, manifest in the heart of all traditions. Muhammad the Messenger, upon him be peace, is President of the Parliament of Prophets. All the members of this Parliament are perfect spiritual equals, sent to every nation in human history with the same essential message of Oneness. Such is the integral view of the Glorious Quran.

Introduction

Four Steps and Seven Levels

My central responsibility in our Dervish Order is to offer initiation and to interpret dreams, which indicate Divine Permission to receive initiation and to advance along the mystic path, characterized by Nureddin Jerrahi by means of twenty-eight Divine Names. Among the four hundred major branches of the Dervish Orders, the path is most often characterized by eight Divine Names, sometimes by twelve, rarely by eighteen. That Pir Nureddin selected twenty-eight indicates that he placed the Divine Seal upon the fullness of the mystic way of Islam.

In the Jerrahi Order, one central initiation offers all the blessings of the path, rather than a series of successive initiations that certain other Orders prescribe. This initiation ceremony is not secret. It is often performed in the presence of visitors to the *tekke,* the dervish meeting hall. I have conducted this rite of entrance and sacrament of spiritual completeness for more than five hundred sincere aspirants, so it has become natural to me, almost like breathing. This ceremony always remains a moving experience for the community as a whole, for myself, for the initiate, and for the mature brothers or sisters who stand on each side of the new dervish, linking arms and helping the aspirant take these four ultimate steps.

The initiation is called *taking hand.* It sacramentally replicates the historical event in the life of the Prophet when certain companions, already loyal to the holy way of life, ceremonially clasped his right hand, marking a vast intensification of their commitment. This act of *taking hand* creates a unique bond with the beloved Muhammad, beyond the respect and loyalty devout Muslims feel for their noble Prophet, upon him be peace. The right hand that is offered and received in this reenactment, therefore, is ultimately the right hand of the Prophet. The right hand of the shaykh is simply a conduit. Out of traditional Islamic courtesy, women initiates do not usually clasp the hand of the shaykh but both hold the same set of prayer beads.

The ceremony is a mystic crowning in which the Crown of Light, usually given to the soul in Paradise, is actually conferred here on earth. Those gifted by Allah with spiritual sight can perceive light, or even a crown of light, descending over the head of the new dervish at the appropriate moment. The Crown of Paradise can be transmitted only in Paradise; therefore Paradise consciousness must become fully present during the initiation. The invisible crown is usually symbolized by the gift of a white cap to the men and a white or colored veil to the women, although many modern women prefer the cap.

Receiving this crown enables one to experience Paradise consciousness here and now, during one's prayers and even during the struggles of daily life. The initiated dervishes can now transmit at least a glimpse of Paradise to their loved ones and colleagues, not verbally but directly, thereby elevating all humankind. The dervishes are not seeking their own spiritual bliss but are clearly motivated by the longing to be of service to humanity and to their own society in particular.

The Shaykh gestures to the experienced dervishes to help the initiate make the first step, beginning with the right foot. The Islamic greetings of peace—*as-salam alaykum, alaykum as-salam*—are exchanged, and the Shaykh welcomes the initiate to the dimension of *sharia,* the depth of the Sacred Law. I welcome aspirants to this exalted level by reminding them that *sharia* is essentially the repetition of the affirmation of Unity, *la ilaha illallah,* externally or internally, verbally or nonverbally, with every breath, every step, every intention, every perception. From this primary pillar of Islam, the other four pillars extend. I remind the aspirant that *sharia* is the way of constant prayerfulness and delight in the prayers, the way of ceaseless acts of generosity and kindness to all beings as one family of consciousness, and the way of fasting—not just abstaining from food and drink from dawn to sunset during Ramadan, but fasting at all times, waking and sleeping, from limited conceptuality and limited emotionality. Finally, *sharia* is the way of holy pilgrimage, but not just to the earthly Kaaba in the noble city of Mecca. *Sharia* is to remain constantly in the open and submitted state of a pilgrim while approaching the true Kaaba, the secret heart of humanity, where the diamond of Divine Essence is concealed from the conventional gaze of the world. This first step, the noble *sharia,* is obviously not just for beginners, nor is it left behind by the next three steps.

The Shaykh beckons the dervish to take another step, and the process is repeated as the aspirant is welcomed to the *tariqa.* This is the steeply ascending path the Holy Quran speaks about, the upward spiraling path that traverses the seven levels of consciousness. This is the path of profound purification, the path of mystic dreams and their inspired interpretation, the path of the joyful uproar and sweet companionship of the dervish lovers of Truth. The *tariqa* is a mystic tree—its spreading roots the beloved Prophet Muhammad, its noble trunk the sublime Ali. The great branches of this tree of Tariqa are the Pirs who have founded initiatory lineages, and the smaller limbs are all the noble shaykhs and shaykhas. The flowers of all colors and fragrances that grow from these branches are the countless dervishes. The fruits are love and wisdom. The sap of this tree is the ecstasy of conscious union with Reality.

The Shaykh gestures again and welcomes the aspirant to the third step, the *haqiqa,* the peak of the mountain of light. Here the path disappears into the boundless green meadow of Truth. Here in Truth alone, the aspirant and the entire community are asked to gaze with the eyes of the heart. Now one can perceive only a shoreless ocean of light—indescribable and inconceivable, without any division or partition, without surface or depth. This ocean of Divine Light is not placid but always filled with giant waves of love. The aspirant is now asked to focus on the eyes of the heart themselves, perceiving that they, too, are composed purely of Divine Light. This is the mystery of *nurun ala nur,* the Light of Allah within the Light of Allah.

The Shaykh beckons a fourth time, and the new dervish takes the final step onto the white sheepskin, laid out in front of the kneeling guide to symbolize the sacrifice of the ego. This is the *marifa,* the courageous descent of the dervish soul from the peak of light into the valley of suffering, struggle, sacrifice, and responsibility, while retaining the conscious union with Truth characteristic of the third step. The culmination of wisdom is to become dust beneath the feet of humanity. *Marifa* is the selfless service of humankind and of creation as a whole, demonstrated by the beloved Jesus, upon him be peace, when he washed the feet of his disciples at the Last Supper, thereby opening their hearts and illumining their minds. The hands of the new dervish now become the Divine Energies, *rahman* and *rahim,* Compassion and Mercy. The heart of the dervish becomes Divine Justice and Divine Love. The breath of the dervish becomes Divine Life. The eyes of the dervish perceive only Divine Beauty. The mind of the dervish operates only with Divine Clarity and by the principle of Divine Unity.

The special protector and guide for *sharia* is the beloved Moses, for *tariqa* the beloved Jesus, for *haqiqa* the beloved Abraham, and for *marifa* the Seal of Messengers, the Distributor of the Light of Prophecy to all Hearts, the beloved Muhammad Mustafa, upon him be peace. A distinct spiritual energy is experienced at each of the four steps. The harmony of all four is ineffably beautiful.

Now the initiate kneels knee-to-knee with the Shaykh, firmly clasping his right hand or prayer beads. The Shaykh prays that the inconceivable Divine Mercy, which is always descending as an invisible rain upon the planetary plane and upon the human heart, should now become visible to the eyes of the heart, cleansing the entire being of the initiate from all misunderstandings or partial understandings imposed since childhood by the limited society or arising from the narrow structures of the limited self. The Shaykh prays that even the slightest shadow of the negation of love should be swept away from this aspiring heart and that it should be filled entirely with Divine Light. Together, the new dervish and the attending senior dervishes, along with the Shaykh and the entire community, repeat eleven times the Arabic phrase *estaghfirullah,* which opens the mind entirely to the power of Divine Forgiveness.

Whenever the Shaykh welcomes a new dervish to the four steps or prays for the aspirant, his words become Divine Energy and bring directly into being, before the eyes of the heart, precisely what is described or prayed, not as an abstraction or as a pious wish but as living Reality. This is the mystery of Divine Creativity described by the Holy Quran. Allah Most High simply calls out the Word of Power, *Be!* and whatever He wills directly and effortlessly comes into being.

At this point in the ancient ceremony of *taking hand,* the Quranic passage describing the original event in the desert of Arabia is melodiously chanted. I interpret the Divine Words to the new dervish in this way. When the lovers of Love linked the right-hand side of their being with the Prophet of Love, upon him be peace, the mystic right hand of Divine Presence descended upon that linking. In this way Allah confirms the original promise made to the noble Adam. This promise has been passed in an unbroken stream of light through 124,000 Prophets to the beloved Muhammad of Arabia and transmitted from him through fourteen centuries of mystic shaykhs. This is the promise of the soul's union with its Lord in the bridal chamber of Divine Love, the promise that even the veils of *soul* and *Lord* will vanish in the supreme realization of identity. Naming the place and year before the eyes of these honorable witnesses, I add that here and now this Divine Promise, which is good until the End of Time, is again being confirmed.

Now the affirmation of Unity, *la ilaha illallah,* is repeated together by Shaykh and aspirant seven times, once for each level of consciousness, the seventh repetition occurring at the level where only Divine Consciousness exists. The Shaykh concludes the seventh affirmation by intoning *muhammad rasulallah,* Muhammad is the Messenger of Allah, and the dervish community begins to sing, in a beautiful traditional melody, the call of Divine Transcendence, *allahu akbar,* the affirmation of Unity and praises of the Prophet. The Shaykh now confers the cap or veil, greeting it thrice with a noble kiss, touching it to eyes and forehead, then offering it to the new dervish to greet in the same manner. The Shaykh places the traditional prayer beads of Islam into the right hand of the fully initiated brother or sister, symbolizing that every breath has now become equivalent to repeating one of the Divine Names. The astonishing fact of initiation is that the dervish has been transformed, before our eyes, into a person of perpetual prayer. His or her individual existence has now become ceaseless Divine Remembrance.

The Shaykh opens his palms and allows words of prayer to stream spontaneously through his heart to his lips. Whatever is appropriate for the initiate is now prayed in a graceful and uplifting manner, precisely as Allah has foreordained. I often conclude this long prayer by supplicating Allah Most High that our Pir Nureddin Jerrahi fix his spiritual gaze upon the heart of the new dervish, night and day, filling it with the Light of Universal Islam, that his saintly mother Amina Taslima transmit her purity and sanctity to this dervish, and that the representative of Pir Nureddin to modern humanity, Muzaffer Ashqi, fill the heart of this dervish with the exquisite wine of Love.

The newly invested dervish kisses the hand of the Shaykh, exactly as if kissing the hand of Pir and of Prophet, stands, and makes the same four steps backward, beginning with the left foot, which symbolizes the mystic way as the right foot symbolizes the sacred law. The atmosphere has now become light, joyous, playful. I reassure the new brother or sister that these four steps backward are not retreat or regression, that none of the spiritual riches of the four steps can be lost, but that one is simply returning to the existential situation, to realize and actualize these sublime gifts that now remain radiant at the core of his or her being. We do not enter the path to engage in religious fantasy but to become more realistic, more free from self-deception, more uncompromising about Truth.

I now request the entire community to embrace the new dervish—or dervishes, for often friends or family members take the four steps together, arms linked in mutual, loving support, hearts merged in the beautiful state of eternal companionship. In traditional Muslim circles, the sisters embrace the sisters and the brothers embrace the brothers, but among North American and Mexican dervishes, these culturally ingrained restrictions often cannot be imposed. After all, the dervishes are one family. There are tears and laughter. The Divine Light shining from the countenance of the newly unveiled dervish is an undeniable, empirical fact.

The most intimate teaching in our Dervish Order comes through spiritual dreams and their inspired interpretation. The shaykh does not deal with psychological or merely stress-releasing dreams, nor is there any fixed system of dream symbolism. Two dervishes came to our previous Grand Shaykh and reported the same dream: climbing a minaret and giving the Call to Prayer. To the first, the inspired interpreter commented, "You are going on pilgrimage. Make preparations." To the second, he remarked, "You have taken something that does not belong to you. Discover what that is and give it back." Before *taking hand* the aspirant often receives a significant dream of Divine Permission or, in some cases, Divine Insistence. After *taking hand* one usually experiences a dream confirming that the ceremony was accepted by Allah. In the context of Islamic spirituality, no sacred rite is considered to be automatically effective. Rather, one must seek and await signs of the Good Pleasure of Allah Most High.

One of the fundamental teachings, shared by the various intertwining lineages of initiation that form the tree of Tariqa, concerns the seven levels of consciousness. Upon this crystal clear analysis of evolutionary levels, the esoteric teachings of Sufism are firmly based.

One does not have to consult ancient textbooks to discover the perennial teaching of Sufism. This esoteric map of consciousness was transmitted with accuracy and clarity in a spiritual dream granted by Allah through the blessings of Pir Nureddin Jerrahi to a Mexican girl of twelve. Along with her mother, father, and younger brother, Rahima had participated in the ceremony of *taking hand* about a year before her extraordinary dream. While visiting Mesquita Maria de la Luz, the Mosque of the Mother of the Prophet in Mexico City, where our Order is led by a gifted and dedicated woman, Amina Taslima al-Jerrahi, I was honored to hear and interpret this dream. In my role as guide, I have listened to thousands of profound dreams during the last eleven years. This one is among the most astonishing. A young girl, with the simple, natural imagery appropriate to her own psyche, accurately pictured the most sophisticated esoteric teaching of Islamic mysticism.

As I listened to her father, Abdul Qadir, translate his daughter's dream from Spanish to English, I began to realize what an immense gift this was to our Order, for we hold in common the spiritual wealth of our dreams and their interpretations. The powerful blessing of a mystic dream does not belong exclusively to the individual dreamer. Its healing, integration, and illumination belong to the entire community. I believe that Rahima's blessed dream of the seven levels of consciousness belongs as well to the lovers of Truth across the whole planet into the distant future.

Rahima dreamed that she was guided by someone she did not recognize through a large house with seven floors. The ground floor was dirt. There were absolutely no signs of human habitation or refinement. The place was not even kept clean. The second floor was an extremely simple dwelling—bare wooden floor, bed, chair, table. It was kept clean and was attractive in its modest way. The third floor was a very comfortable home, according to modern standards. There were carpets, radio, television, refrigerator, and so forth.

When Rahima was taken to the fourth floor, the fourth level of consciousness, she was amazed to find a brilliant palace—marble floors, high ceilings, large gilded mirrors, beautiful antique furniture, precious ancient vases, and other works of art. At this point in the recounting of the dream, I began to realize that certain mysteries of the spiritual path, which remained vague to me, were about to be displayed in simple, dramatic imagery. All who were present entered a mild state of ecstasy, a gift of the fourth level. Rahima continued speaking, calmly and confidently, without any self-consciousness.

When the dreamer was guided to the fifth floor, she encountered total darkness, filled with a deep, rumbling music that she, as a twelve year old, found rather unsettling. When taken to the sixth floor, she found an empty, candle-lit space where a circle of dervishes, wearing white and kneeling on sheepskins, were engaged in the ancient ceremony of Divine Remembrance.

Arriving at the seventh floor, Rahima entered a brilliant, sunlit room, illuminated through large skylights and filled with lush green plants. No person was present, nor were there indications of human habitation. The golden light and the dark green of the leaves created a joyful, expansive feeling. Suddenly, one of the plants reached toward her with a long creeper, wrapped around her waist, and gently threw her out an open window. She fell with equal gentleness to the earth below, landing on her feet.

Almost as an afterthought, Rahima mentioned that her guide took her back through the same sevenfold structure several times, so that she was perfectly clear about the various levels. Each time, she was thrown out the window again. I asked her how many times she ascended these floors. She thought carefully for a moment, then replied definitively, "Four times."

The interpretation of this dream can be extensive. I offered a seminar in Mexico City on the seven levels of consciousness, during which I spoke about this dream for several hours.

The first level is the domineering self, basis for the aggressiveness, territoriality, and violent urge for survival that seriously threaten the coherence of our personhood, our society, and our planet. There is nothing intrinsically human here. There is no possibility for hospitality. There is not even the cleanliness that is essential for human dignity. Although most human beings experience disconcerting flashes of this domineering ego, very few persons remain focused on this level. Only war criminals and other enemies of humanity could be said to live primarily on the first level of consciousness. Nevertheless, there is nothing intrinsically evil about this first level. It provides a biological ground floor for human reality. Through this consciousness, the lungs breathe and the heart beats.

The second floor in the dream represents the critical or inquiring self. Most of humanity is focused on this level, where basic human refinements are beginning to appear. This dream imagery has nothing to do with social standing or affluence. There are persons living in presidential palaces who are occupying the dirt floor of the first level of consciousness, as well as persons who live in thatched huts who are enjoying the glorious palace of the fourth level of consciousness.

The evolutionary efforts carried on by this second level of the self constitute the critique of the domineering ego, the critique of selfish impulses. The search is carried on here for truly human and humane values, for disciplined and fruitful ways of life. There are many dimensions within this second level of consciousness. They are all essentially positive, honorable, and evolutionary, unless they remain dominated by the first level, obviously or subtly.

The third floor of this structure of consciousness is the fulfillment of our humanity. Human potential is here unfolded harmoniously. Perhaps the majority of human beings reach upper regions of the second level, but only excellent persons of good will become established on the third level. Here, ethical and religious ideals are in full flower. This level of development, or awakening to our true nature, is the real basis for civilization—religion, education, art, science. Sincere seekers on the second level receive certain glimpses of the third level, but where one's awareness remains primarily focused is what counts for evolutionary development. In traditional Sufi parlance, the third level is the fulfilled or satisfied self.

One could reasonably inquire, how can there be levels higher than this fulfillment of human aspiration to an excellent, civilized existence? The four higher levels are the fruition of the mystic path of return. They are not, strictly speaking, part of human potential and human effort. They are the manifestation of Divine Reality through our human reality.

One usually must reach the third level of consciousness to receive authentic initiation into a mystical Order, or one may be lifted by Divine Grace, through this initiation, into the third level. When one reaches the fourth level, Divine Attributes begin to manifest directly and adorn the human being. This is symbolized in Rahima's dream as rare works of craftsmanship and art. These manifestations are not, however, works of human hands, nor are they brought about by human efforts. The transition to the fourth level usually occurs after physical death in the realm of Paradise consciousness. Only genuine mystics can generate enough spiritual intensity to enter this and higher levels during earthly experience. Once again, we recognize that gifted persons on the third level, or even on the second level, may receive glorious intimations of the fourth level of consciousness, but to be established there is an entirely different order of experience. Not even all the members of a mystical Order become established on the fourth level, which in traditional Sufi parlance is the tranquil self.

The fifth level is that of mystic union, where no finite modes of thought or perception operate, hence the symbol of total darkness. The thunderous music in the dream represents the Divine Resonance, from which universes are taking shape and into which finite existence disappears again. This was the only floor in the dream structure that caused Rahima nervousness and concern, since this radiant blackness is so far from our ordinary level of experience. In Sufi parlance, the fifth level is the peaceful self.

If we were to correlate the seven levels with the four steps, *sharia* would be the third level, *tariqa* the fourth, and *haqiqa* the fifth. The final two levels of consciousness are an expression of *marifa,* the astonishing dimension of spiritual manifestation that lies beyond mystic union. On the fifth level, there is only Truth and its Resonance. On the sixth level, creation appears once more, not through beautiful Divine Manifestations, as on the fourth level, but as the mystic crown, the sublime human form, symbolized by the circle of dervishes. One surprising piece of good news brought by Rahima's dream is the confirmation that the ancient ceremony of *dhikr,* traditionally conducted by candlelight, kneeling on sheepskins, actually affords the blessed dervishes in the circle a glimpse of the sixth level, although most of them may not even have become established on the fourth level. In the precious sacrament of *dhikr,* essential Divine Energies descend through the hearts and even through the physical bodies of the dervishes. Divine Reality becomes visible and experienceable as human reality. In Sufi terminology, the sixth level is the complete self.

The enigmatic seventh level of consciousness is a realm of brightness, clarity, subtle humor. The human form has been transcended, even as a mode of pure Divine Expression. Thus the seventh level resembles the fifth level in its absence of human reference. Yet here the imagery of light and luxurious growth replaces the imagery of mystic darkness. The human person of Rahima was not permitted to remain but was removed instantly in a playful and humorous manner. My Shaykh, Muzaffer Ashqi, used to comment simply, "On the seventh level of consciousness, if you imagine that you exist, it is idolatry." By the dynamic golden greenness of Supreme Reality, all possibility of the idolatrous perception of duality is tossed out the window. The colors on this seventh level indicate why Nureddin Jerrahi designated a golden cap wrapped in green cloth as the turban of his Order. Green is also the chosen color of the beloved Messenger of Allah. In Sufi parlance, the seventh level is the pure self.

Rahima was taken through this symbolic dream structure four times, indicating that she, although only twelve years old, was already in communion with the fourth level of consciousness. As she grows older, she will have to practice spiritual discipline and experience intense yearning to become fully established on this fourth level and to progress further. This dream is itself one of those rare works of Divine Art that manifested in the palatial fourth floor of her dream. Her unknown guide was probably Nureddin Jerrahi, may his spirit be sanctified, whose intercessory power, by the Permission and Foreknowledge of Allah, tenderly opened the way for this amazing dream, which has now become a channel of spiritual energy and illumination for us all, her grateful brothers and sisters.

PART ONE

Traditional Islamic Resources

AFFIRMATION OF UNITY This metaphysical poem was composed in 1981 during the long flight from New York City to Istanbul. It was a spontaneous offering to my Shaykh, Muzaffer Ashqi, and my Pir, Nureddin Jerrahi. I can see this document now as confirming the completion of my apprenticeship. It focuses on the highest teachings of Islam concerning the preeternal Muhammad of Light, teachings comparable in tone to the Christian mystical doctrine of Logos. This poem is not based on personal reading or speculation but on oral transmissions and heart transmissions I received from my Shaykh during three years of discipleship and friendship, beginning in 1978, which included our pilgrimage together to Mecca and Medina in 1980. This poem presents rich instruction in the basic contemplative exercise of the dervishes: the affirmation of Unity, *la ilaha illallah,* and the affirmation of humanity, *muhammad rasulallah*.

Rasulallah is used instead of the more accurate English transliteration *rasulullah* in order to preserve the name *Allah* for the understanding of persons not familiar with Arabic. Similar decisions on the English transliteration of Arabic phrases have been made throughout this book.

1
Affirmation of Unity
la ilaha illallah muhammad rasulallah

With each breath may we take refuge
in living Truth alone,
released from coarse arrogance and subtle pride.

May every thought and action be intended,
in the supremely holy Name of Allah,
as the direct expression
of boundless Compassion
and most tender Love.

May the exultation of endless praise,
arising spontaneously as the life of countless beings,
flow consciously toward the single Source of Being,
Source of the intricate evolution of countless worlds.

May we be guided through every experience
along the Direct Path of Love
that leads from the human heart into the Divine Heart,
the ever-present Source of Love.

With beautiful Names of the All-Merciful One,
Holy Quran reveals the secret of infinite Mercy:
Allah is the perfect Oneness and utter Completeness
that embraces every world and every being.
The Supreme Source that calls Itself Allah
and by countless other Beautiful Names

has not come into being from anywhere,
nor can any being come into being
separate from the Source of Being.
There is only One Reality.
Planetary and heavenly realms
and the countless beings they contain
are simply the Attributes of Allah
praising the Essence of Allah.
This is the Truth revealed by Allah
within the heart of humanity.

la ilaha illallah
There is no reality apart from Ultimate Reality.
Allah alone is worthy of worship,
for Allah alone is.

la ilaha illallah
As Allah Most Resplendent
reveals to the beloved Moses,
"I alone am.
There is absolutely nothing
apart from the boundless
I Am that I am."

la ilaha illallah muhammad rasulallah
Divine Unity alone exists,
and humanity is
Its principle of Self-revelation.

la ilaha illallah muhammad rasulallah
The Mercy of Allah to the universe,
the beloved Muhammad,
may Divine Peace embrace him always,
discloses to his spiritual companions throughout time,
"Allah is the Hidden Treasure
Who longs to be known
by the intimate knowing that is love."

The longing to be revealed
arises spontaneously
within the mysterious hiddenness
of sheer Transcendence,
without disturbing Its perfect Unity.
This Divine Longing is the First Light
to emerge from the Source of Light, *nur muhammad,*
the exalted Muhammad of Light,
principle of Divine Manifestation,
Light of Guidance and Light of Prophethood.
This primordial Light from light
shines prior to time and prior to eternity,
prior to the existence of any realm.
The *nur muhammad* gazes everywhere,
perceives only Divine Unity
illuminated by Its Own Light,
and calls out *la ilaha illallah:*
There is nothing other than Allah.
Allah Most Resplendent and Sublime,
Whose mystic hiddenness now stands revealed,
responds with Divine Delight,
muhammad rasulallah:
O Muhammad of Light,
you are My Principle of Revelation.
The boundless Universe of Souls
and the eighteen thousand dimensions
where souls manifest
Allah creates only through and for
His beloved Muhammad of Light.

la ilaha illallah
Turning away from the heart,
the dervish chants *la ilaha,*
there is nothing apart—
no limited world, no limited self,
no limited principles, no limited powers.
la ilaha empties the universe

so manifest Being becomes translucent.
Then the dervish turns toward the physical heart,
which orients the inward gaze toward the spiritual heart,
and experiences the spontaneous affirmation
illallah, there is only Allah.
illallah fills the perfectly empty vessel
with the radiance of Supreme Reality.
Wherever we look, Truth alone is shining.
Within the profound resonance of *illallah,*
the dervish hears the Divine Response
emerging from the inmost heart,
muhammad rasulallah,
for springing forth eternally within Divine Unity
is the secret exaltation of humanity
as the complete expression of Divine Love.

The chanting of *la ilaha illallah muhammad rasulallah*
is none other than the primordial hiddenness of Allah
revealing Itself to Its Own beloved Light.
Through the sanctified voice of the dervish,
the Muhammad of Light affirms, *la ilaha illallah.*
Through the secret heart of the dervish,
Allah Most High responds, *muhammad rasulallah.*
There is nothing else and no one else.

la ilaha illallah is sharp and clear as a diamond,
muhammad rasulallah delicate and fragrant as a rose,
both revealing only One Reality.

la ilaha illallah is Mecca, Sun of Divine Power.
muhammad rasulallah is Medina, Moon of Tender Love.
Mecca strips the heart naked before Truth,
Medina clothes it in silken robes of ecstasy,
both revealing only One Reality.

la ilaha illallah is the Resplendent Unity beyond imagining
for Whom even the sublimity of Paradise is a dream.

muhammad rasulallah is the universal Light of Guidance,
streaming through 124,000 beloved Prophets of Allah,
that illuminates the mystic path of Love
along which souls of Love return
to disappear into the Source of Love.

The radiant Drama of Love, *muhammad rasulallah,*
unfolds entirely within Divine Unity, *la ilaha illallah,*
for nothing exists outside this perfect Unity
and nothing can disturb Unity from within.

la ilaha illallah is the constant dissolution
of limited worlds and limited selves
into the radiance of limitless Truth.
muhammad rasulallah is the exaltation of the soul,
whose essence is Truth,
for Allah Most High decrees that every being
on the higher planes of Being
prostrate before the transcendental Adam,
archetype of the precious human soul,
and none can bow except before Truth.
All existence loses itself and disappears into Allah
except the unique diamond of the soul
that remains with Allah in eternal companionship
as the mystery of *nurun ala nur,*
the Light of Allah within the Light of Allah,
manifest through the companions of Love
who gather every Friday in the realm of Paradise
for the dervish celebration of Divine Love.
The exalted Muhammad of Light
takes the position of Shaykh,
124,000 radiant Prophets of Allah form the inner circle,
intimate friends and slaves of Allah the second circle,
humble servants of those slaves of Love the third circle,
souls loyally devoted to the holy way of life
revealed through the Messengers sent to every nation
compose the vast outer circle of chanting dervishes,

and those who love any of these beloved ones
are gathered mercifully by the All-Merciful One
to experience the sweetness of Divine Remembrance.
These luminous souls are singing in perfect unison
la ilaha illallah muhammad rasulallah,
lost in the mystic moment beyond time and eternity
when the Hidden Treasure
stands revealed by Its Own First Light
as the incomparable Treasure of Love.

la ilaha illallah is the brilliant blue sky of Truth,
muhammad rasulallah the graceful dark rain clouds,
the eighteen thousand dimensions of Divine Love's Drama.
How can the radiant Storm of Love
gather in the boundless Sky of Clarity?
Why does Allah Most High exclaim, *Be!*
so that countless forms of consciousness
come into being on every plane of Being?
Why does infinite Mercy guide some
along the Direct Path of Return to the Source
and others along paths that wander among shadows,
turning away from the Source of Light?
What is the meaning of Allah's Revelation
in the Holy Quran
that by each thought and action without exception
every living being remembers the Source of Being?

la ilaha illallah is the clear gem with infinite facets,
goal of every profound quest,
response of Truth to every ultimate question,
principle of utmost clarity
that leads beyond every concept and description,
healing of every doubt
with the sweet balm of surrender,
total illumination of mind and heart
that dissolves every cell of the body
and every atom of the universe into Light.

muhammad rasulallah is the Divine Drama,
the vast Circle of Revelation that begins
with the exalted Muhammad of Light
and reaches completion through
the beloved Muhammad of Arabia,
may the sublime Peace of Allah embrace him,
the Culmination of Prophethood
through whom the Light of Guidance
now streams into every human heart
as the fullness of Allah's Mercy.
Promised through Jesus
as the Comforter and Friend of all souls,
the beloved one of Allah served
as diamond vessel for an Arabic Quran
that would crush vast mountains
with its weight and power,
emanating from the Transcendent Quran
where every personal and cosmic event is written.
By this radiant and living Book of Reality
the profound meaning of all events
will be perfectly illuminated
on the Day of Truth when time ends.
The entire Drama of Revelation,
from Muhammad the Principle
to Muhammad the Beloved,
emerges spontaneously from *la ilaha illallah*.
The Divine Drama, *muhammad rasulallah*,
is simply *la ilaha illallah*
unveiled as the Treasure of Love.
There is only *la ilaha illallah*.

la ilaha illallah To be is to praise Allah.
The inner function of manifest Being,
including planetary and heavenly realms,
is simply to praise the Source of Being
and to be lost in exultation.
To be is to be returning

constantly to Allah.
To be is to disappear into Allah.

muhammad rasulallah To be human
is to lose the limited self in love,
to live only for love,
to engage continuously in acts
of profound compassion and common kindness
toward all beings as one family of consciousness,
dedicating each action
to the principles of justice
and reverence for life revealed by Allah,
feeding the hungry on every level of hunger,
caring intensely for all creations of Allah
through whom one Divine Life is pouring.

la ilaha illallah The radiant kingdom
of the infinitely merciful and responsive King,
always present to every subject from within,
is the intricate and harmonious display
of eight planes of Being,
where no soul is wronged
by even so much
as the point of a date stone,
perfect training ground
for the human heart.

la ilaha illallah muhammad rasulallah
Expressed by formless splendor
and by the beauty of form,
there can only be
one inconceivable Reality.

la ilaha illallah muhammad rasulallah
While chanting the powerful affirmation *illallah,*
the dervish gazes inwardly toward the true Kaaba,
the precious human heart,

whose secret is *muhammad rasulallah.*
The marvelous organic heart that ceaselessly
remembers Allah with its pulse of life
forms the outer gateway.
The heart of tender emotion
contains the royal garden.
Tracing the essence of awareness there
among flowers of radiance and fountains of resonance,
the dervish comes upon the sublime spiritual heart,
the translucent emerald palace of the King.
Overwhelmed and drawn further
by the sweet rose fragrance,
muhammad rasulallah,
the dervish opens mystical doors
with the master key of *illallah*
and passes through eleven concentric courts,
each one more radiant with Divine Beauty.
They contain eighteen thousand universes,
seven progressively more subtle heavenly realms,
and ninety-nine levels of Paradise.
The dervish now discovers
that whatever appears to exist
outside the heart
is mere reflection.
At last the inmost door of the secret heart,
touched by the powerful key of *illallah,*
swings open to reveal the Throne of Light,
surrounded by royal ministers of Divine Love,
moving in wise attendance and graceful praise.
One final threshold beckons,
the luminous open portal
to the private chamber of the Sultan.
Entering there, the dervish disappears.
Only the Muhammad of Light remains,
forever in communion with the Sovereign,
revealing the Sovereign, knowing the Sovereign,
eternally lost and eternally found in Divine Love.

Here is the Essence of Love,
transmitted from the supreme lover
through his lineage of mystic guides
to Sultan Muhammad Nureddin Jerrahi,
beloved of the beloved one of Allah,
whose love streams into the world
through the entire being of those
who are always deeply in love
and who belong only to love.

la ilaha illallah Chanting dervishes
are the melodious thunder of Allah,
waves of love in the stormy ocean of Love,
breaking in ecstasy against the Throne of Allah.
The heart of the dervish is a tiny golden ball
cast high into the boundless sky of Truth
by immense waves of Divine Resonance.
Who exists other than Allah
to affirm that Allah alone exists?

la ilaha illallah is Allah's Own Remembrance.
From the power of this living Word
the waves of worlds and beings
are constantly flowing and evolving,
and into the formless Source of Sound
they race in delight to disappear again.

la ilaha illallah Deep within the thundering Word,
the dervish encounters the essential silence
where the drama of time and eternity ends,
the Day of Truth toward which creation is rushing,
which shines already as the secret of secrets,
as brilliant silence hidden within the Resonance of Allah.
From this essential silence,
the eighteen thousand universes have never gone forth.
From this silent Essence,
the most precious secret soul

has never been separated.
hu hu hu O Hidden Treasure,
eternally revealing, yet hidden still.

muhammad rasulallah
O First Light, Beloved One of Allah,
Perfect Soul, *nur muhammad,*
moon of tenderness, forever full,
sun of knowledge that never rises or sets,
consummate spiritual guide,
Shaykh of shaykhs.
May the sublime Peace of Allah,
the exaltation of Paradise,
and the disappearance into Divine Love
belong to you always, to those who love you,
to those who transmit your guidance,
and to those who love them,
O most beloved Muhammad of Light.

muhammad rasulallah
O diamond essence of the soul,
may your sword of light,
the principle *la ilaha illallah,*
reveal universal Islam
as the sublime peace of Unity,
hidden in the heart of all sacred traditions,
dawning in the heart of every human being,
shining in the heart of manifest Being
that constantly praises the Source of Being.

muhammad rasulallah
O trustworthy guide and friend of souls,
the precious Light of Guidance streams through you
as 124,000 Messengers of Truth.
O Muhammad of Light, standing eternally
on the Mountain of Light above the City of Light,
dissolving the universe entirely into Light

with your merciful Sword of Light
held high in the stillness of Truth.

Most profound salutations of the heart
to beloved Abraham, to beloved Moses, to beloved Jesus,
and to the beloved Muhammad of Arabia,
may supernal peace embrace them all.

Most tender greetings of the heart
to beloved Virgin Mary, mother of Jesus,
pure channel of Divine Love for all humanity.
Greetings to noble mother Khadija,
first follower of the Prophet
and protector of Islam,
to precious Aisha,
most intimate wife of the Prophet
and perfection of womanhood,
to cherished Fatima,
radiant daughter of the Prophet
and exalted Mother of the Faithful,
to magnificent Ali, Golden Lion of Allah,
instrument chosen by the Prophet
to establish his mystic path,
the way of union with Truth,
to most revered Hassan and Hussain,
grandchildren of the Prophet,
bearers of the Light of Islam.

Most profound salams to the Prophets
sent by Allah All-Merciful to every nation,
kisses of loving respect
to the hands and feet of the holy shaykhs
who inherit the spiritual wealth of the Prophets,
most loyal friendship of the heart
to all lovers of Truth,
known to the world or hidden from the world.
May we be accepted into the radiant circle

of eternal Divine Remembrance
formed by the highest companions of Love.

O Allah Most High,
Cherisher and Sustainer of all worlds and beings,
please accept every breath as prayer,
as the constant offering and surrender
of our individual lives to the Source of Life.

ya hayy, Thou All Life.
ya nur, Thou All Light.
ya aziz, Thou All-Powerful Love.
ya salam, Thou All Peace.
May we consciously abide
in the universal Islam
of Your Peace.
amin

SALAT This essay may be unique in modern Islamic literature. The five times daily prayer, or *salat,* is often presented in a mechanical, technical fashion, emphasizing the sense of strict obligation. The present composition does not pretend to be a manual for the practice of Islamic prayer, which is an important literary genre, but aspires instead, by drawing upon the most profound and beautiful Oral Traditions of the Prophet, to introduce the sublimity of this practice, this complete way of life, which sanctifies temporality. Like the poem preceding it, this essay is not the fruit of reading or speculation but reflects verbal and nonverbal transmission from my Shaykh and his prayerful community of dervishes. Muslims representing every education and cultural background will attest to the authenticity of this presentation, which is not exhaustive but which is infused by the incomparable rose fragrance of *salat*.

2
Salat
Daily Prayers of Islam

In the Name of Allah, the Merciful, the Compassionate

The sublime Oral Tradition, or *hadith*, reports that Prophet Muhammad, upon him be peace, received *salat*, the five times daily prayer of Islam, during his mystical ascension through the heavens and beyond Paradise into the glorious Garden of Essence. During this journey, the Messenger of Allah was able to contemplate angels performing each of the various movements of *salat*. Some stood in rapture before the Divine Majesty, others bowed continuously in astonishment and awe, still others remained lost in full prostration—all of them merged consciously in Divine Unity. Therefore, *salat* is a gift from Allah, displayed first through angelic beings, rather than a ceremonial form springing from human intellect, will, or initiative. The Prophet of Islam proclaimed, "*Salat* is the ascension of the faithful," thereby indicating that his own mystical experience of ascending through the heavens to the Divine Throne and beyond can be replicated in the life of the humble practitioner of daily prayer.

Ascension culminates in mystic union. *Salat* is more a way of union with God than an offering to God. The Divine Completeness does not ask for or need any offering. Since Allah is the sole source of action in the universe, the Divine Power is performing prayers as well as receiving prayers. This is why the intimate lovers of Allah, while engaging in the traditional movements of *salat*, experience the energy to bow, prostrate, and stand again as coming directly from the Source of the Universe. *Salat* exists beyond the kingdom of personal will. It is the key to the Kingdom of Divine Will.

Muhammad, the beloved one of Allah, could have brought back from his ascension any gift from the infinite Divine Treasury. Since he returned with *salat,* we can infer that it is most precious.

As the noble Prophet, upon him be peace, explained in his concise manner, "*Salat* is the best *dhikrullah,*" referring to *dhikr,* the practice of constant inward remembrance of Allah. Therefore, the Sufis, or mystics of Islam, prepare themselves carefully through profound spiritual exercises, or *dhikr,* in order to experience *salat* truly. The daily prayers of Islam are not just for beginners along the holy way but also constitute the experience of culmination for those most advanced in the path.

Our Grand Shaykh Muzaffer Ashqi al-Jerrahi, may Allah sanctify his spirit, breathed his last breath in prostration, forehead resting upon his prayer carpet, while performing the supplementary prayers of midnight. A Master of Essence thus merges with Divine Essence during *salat.* An Oral Tradition of the Prophet confirms that to pray five times every day is equal in spiritual intensity to praying fifty times a day. In other words, *salat* leads to continuous prayer. Its goal is the perpetual inward recollection of Supreme Reality.

The beloved one of Allah explained to his blessed companions, "Each step you take while approaching the mosque to pray is counted by Allah as pure prayer." He also indicated that the time during which one remains awake, waiting for night prayer, is also counted by Allah Most High as prayer. In countless ways, the peaceful atmosphere and transforming power of *salat* overflow the five formal periods of prayer, permeating every moment of awareness with holy expectation.

Allah reveals to His beloved Prophet in the Glorious Quran that his principle task as a human being is prayer. This is why *salat* was so deeply cherished by the noble Muhammad, upon him be peace. He proclaimed, "Allah Most High has caused me to love three above all: women, perfume, and prayer," indicating that to his refined spiritual sensibility, prayer is continuous with the delight and fragrance of love and beauty. This transcendent pleasure, this spiritual delight that is also sensual, is felt at least during certain occasions of prayer by every faithful person who practices *salat*.

The central longing of every Muslim man and woman is to live in exactly the same spirit in which the noble Prophet lived—that is, to follow his *sunna*, his most beautiful model, both in attitudes and actions. Delight in prayer is the most important *sunna*, which is linked with the other principle prophetic characteristic, ceaseless expression of kindness, affection, and justice toward all living beings. Without immersion in prayer, universal compassion is not possible.

For Muslims, *salat* is ordained by Allah as the most effective way to unfold the fullness of our humanity. Allah does not request or need our prayers but offers *salat* to us as His supreme gift. The Muslim prays not from a sense of obligation but as an act of tender responsiveness, as a lover desires to fulfill every wish of the beloved. It is not to the Prophet or for the Prophet that we pray, yet the fact remains that his prayer has become our prayer. We are not praying to please him; we are praying with him and as him.

The beautiful form and precise timing of *salat* are the way the noble Muhammad prayed. Yet his life was so filled with praying that both the basic *salat* and the supplementary *salat*, which constitute forty cycles of prostration each day, represent only a drop from the ocean of his prayer.

The beloved Aisha, may her soul be ennobled, the youngest wife of the holy Prophet and a woman who achieved the exalted rank of Mother of the Faithful, related that frequently during the deep night her noble husband stood for prayer, right on their bed, which was composed of a thin leather mattress. When Aisha was resting between him and the sacred city of Mecca, she would withdraw her legs as he bowed, giving him enough space to place his luminous forehead in prostration. She would extend her legs again when the Prophet stood up to chant long passages from the Holy Quran.

By contrast, whenever the beloved Prophet directed *salat* in congregation, he would seldom chant very extensive Quranic passages because, as he remarked on one occasion, "I can hear the crying of small children, and I do not want to cause more difficulty for their mothers." In certain instances when leading the communal prayer, the Prophet of Islam would lose himself in ecstasy and pray many more cycles, or *raqats,* than usual. Afterward, his loyal companions would inquire, "O Messenger of Allah, has a new form of *salat* been revealed?"—so open were these companions to the ever-unfolding Will of Allah for their community and its leader! The Prophet would respond, "No, I simply became lost in the delight of prayer."

Recorded by Oral Tradition, the beloved one of Allah proclaimed, revealing secret teaching from the Most High, that *salat* performed in congregation is twenty-seven times more powerful than the very same prayer completed in solitude. This is the source of the longing to pray in congregation that every Muslim man and woman feels, although solitary *salat* is spiritually valid and sufficient.

The unified lines of prayer, shoulders and feet touching, with the intention of the hearts in perfect alignment as well, produce such an experience of spiritual power and ecstatic communion, it is as though all humanity were standing together in *salat.* The lines are carefully kept straight to create human unity, which then becomes a vessel for the consciousness of Divine Unity. Often the Prophet of Allah would be seen walking among the lines of the faithful, meticulously straightening them before beginning the process of *salat,* during which individual differences melt away and humanity prays as a single body, a single consciousness. *Salat* is the blissful merging of our small personal will with the encompassing Divine Will.

Were these straight lines of prayer to be extended for thousands of miles, they would begin to curve and eventually form planetary circles around the earthly axis of prayer, the holy Kaaba in the ennobled city of Mecca. Thus *salat* at the Grand Mosque in the desert of Arabia is visibly performed in concentric circles around this sacred cubic structure, originally built there by the Prophet Abraham, spiritual father of Jews, Christians, and Muslims.

The very nature of the human soul is prayer, even if we remain largely unconscious of this fundamental level of our being. The Holy Quran reveals that when birds open their wings for flight, this movement is a form of prayer. Allah is continuously aware of the distinctive forms of prayer flowing naturally from the awareness of all creatures. Even stars and atoms whirl in universal praise. When standing for *salat,* one directly and tangibly faces the oceanic radiance called Allah. Noble Ali, leader of the mystics of Islam, used to proclaim that he would not pray to any divinity he could not experience directly. To prostrate during *salat* is to plunge directly into the Ocean of Allah.

The Glorious Quran reveals that all souls were created in the preeternal universe of souls, before planetary or even heavenly realms existed. The human soul is therefore more ancient and more sublime than the angels, more original even than the Divine Throne, which the angels circle in ecstatic praise.

Allah Most High asked all the timeless human souls who were to manifest in time whether they wished to return consciously into their original Source. All souls without exception, including our own, responded in perfect unison, "Yes. You alone are our precious Lord." *Salat* is a clear manifestation of that *original yes,* more original than what Christianity terms *original sin* and Islam describes as the descent of the prophetic souls, Adam and Eve, into the veils of space and time to experience mystic longing for reunion with their Lord. Allah Most High reveals in His Glorious Quran that Being is created to function solely by praising the Source of Being. To be, therefore, is to pray or praise. The gift of *salat* is to experience this astonishing fact continuously, moment by moment.

The Quran repeatedly reveals that Allah does not need, demand, or compel the prayers of any conscious being, for Allah is perfect Completeness. The mystical commentators explain that Allah does not need the prayers of any being because apart from Allah no being exists. The entire creation is simply the Attributes of Allah praising the Essence of Allah.

The graceful movements of *salat,* the careful reckoning of the exact times for *salat,* the profound Quranic learning and melodious Quranic chanting useful for the performance of *salat,* the powerful spiritual affirmations that punctuate *salat*—all these are the golden setting of the ring. The priceless diamond in this setting, the very essence of *salat,* is the revelatory Arabic of the Quran—Divine Word, Divine Resonance, Divine Radiance. These Quranic verses are definitely not human words, although some of their infinite meanings are comprehensible to the human mind by the Permission of Allah. During the external or internal Quranic chanting of *salat,* it becomes existentially clear that Allah alone exists, both praying and receiving the prayers. Since the Prophet Muhammad, upon him be peace, was the human vessel for the Arabic Quran and is even referred to in Islamic tradition as the Quran in human form, we can conclude that the noble Prophet is *salat.* We are praying as him as well as praying within his prophetic heart.

During his last day on earth, the Messenger of Allah, physically weak, came forth from his simple desert abode to gaze upon his community as they performed the congregational prayer. Abu Bakr, may his soul be ennobled, first successor of the Prophet, was leading *salat* but moved to one side when he noticed the noble Muhammad standing in his doorway. Abu Bakr humbly invited the Messenger to take his proper station as *imam,* or leader of the prayer. The beloved one of Allah responded with a silent gesture, indicating that his dear companion should continue to lead *salat.*

An eyewitness reported that while the Prophet stood in the doorway of his humble house, enjoying the beautiful sight of the congregational prayer, his countenance shone like a golden Quran, like a rising sun on the horizon of planetary spirituality. That very afternoon, he passed away from the visible world into Divine Beauty, knowing that *salat* had been truly established upon the earth and within the human heart, knowing that his task had been completed. The Prophet could never have died before attaining certainty that his sole function, continuous prayer, had been successfully transmitted to his community and to planetary history.

Just as the golden sun continues to rise above the horizon, the early morning Call to Prayer flows in never-ceasing waves around the globe, followed by four other waves of *salat* within the time of a single earthly revolution about the solar axis. The noble Imam of all Muslims, the leader of *salat* for the entire world, stands outside time in his transcendental aspect as the Light of Guidance. He faces the prayer niche, or *mihrab,* in the Grand Mosque of Medina, which now surrounds the holy tomb, the fragrant resting place of his external form in the exact location where the Prophet watched *salat* for the last time with physical eyes.

It is behind this Muhammad of Light, and behind no other, that Muslims of fourteen centuries have faithfully continued to form the lines of *salat.* The Grand Mosque at Medina is the human heart.

Call to Prayer

The first moment of *salat* is the Call to Prayer, the melodious cry that opens the hearts and souls of Muslims and non-Muslims alike. Why is the Call so powerful? Because it is not simply a human phenomenon. Supreme Reality is calling us to *salat* through the beautiful human voice. This Divine Call is not merely an announcement, indicating that one of the five periods of prayer has arrived. It is an astonishing invitation to enter the audience chamber of the Sovereign. More amazing still, it is the Sultan of Being who is calling to His beloved souls to pass beyond the outer chamber and to enter mystic communion and union with His Essence. We can long to commune with Allah Most High only because Allah longs to commune with us. *Salat* is the expression of this Divine Longing.

The Call to Prayer is such an integral part of *salat* that even in solitude one does not omit it, unless one has already heard it from a distance. This Call from Allah indicates that each invisible wave of *salat* emerges from and returns to the Most High, simply flowing through the visible human instrument, blessing humanity and all creation as it passes. The fact that one cannot begin *salat* before the precisely prescribed sun time or without the Divine Call gives further evidence that the prayers of Islam are not to be considered human initiative. Allah is reaching out to us, through us from within, not we to Him.

The times for beginning *salat* are Allah's, not ours. For this reason, to make *salat* at the exact moment the particular prayer period arrives allows us to experience its sacramental plenitude, for then we are accepting prayer as the Divine Decision, not as our personal decision, however well intentioned. Naturally, personal prayer is appropriate in almost every moment, in almost every place. But divinely ordained *salat* is the unshakable basis upon which personal prayer rests, not the other way around.

While reflecting on some of the meanings of this primal Divine Call, heard in one form or another by all the Prophets during human history, we should remember that the words of the Call to Prayer are more than meaning. By hearing these words, one is penetrated and transformed by their initiatory power and is lifted, every time, into their atmosphere of spiritual grandeur.

> *allahu akbar* (four times)
> Supreme Reality is always greater than any conception
>
> *ashhadu anna la ilaha illallah* (twice)
> I witness that there is no reality apart from Supreme Reality
>
> *ashhadu anna muhammadan rasulullah* (twice)
> I witness that Muhammad is an authentic Messenger of Supreme Reality
>
> *hayya-s-salat* (twice)
> Come to *salat*
>
> *hayya-l-falah* (twice)
> Come to the highest spiritual realization
>
> *allahu akbar* (twice)
> Allah is always greater
>
> *la ilaha illallah*
> Nothing exists apart from Allah

The opening words, *allahu akbar,* signify that Allah, the sole Reality, is always greater than any possible human conception or imagination. This astonishing exclamation is repeated four times to create the fullest intensity of insight, and it reappears throughout the graceful movements of *salat* to reinforce our awareness of this fundamental fact—Divine Ineffability.

At the second cry, *ashhadu anna la ilaha illallah,* we become witnesses to the indescribable grandeur of Divine Unity, witnessing that nothing can exist outside this indivisible, essential grandeur. There is no other. The affirmation is repeated twice because it does not consist just of verbal meaning but of healing, purifying, and illuminating Divine Energy. *Ashhadu anna la ilaha illallah* recurs at the central point of *salat,* opening the door to the direct experience of Supreme Reality. It is not surprising that a condensed form of this affirmation, *la ilaha illallah,* is most favored by mystics of Islam in both their communal and solitary spiritual practices, called *dhikr.* The Holy Quran reveals that not just mature human beings and angelic beings but Allah Most High stands as witness to His Own transcendent Greatness. The human witnesses, and the entire creation, are simply participants in the Divine Witnessing, the Divine *I Am.*

The third cry, *ashhadu anna muhammadan rasulullah,* refers not only to the noble Muhammad, upon him be peace, but to the entire prophetic lineage beginning with the noble Adam, Prophets who the Quran confirms have been sent to every spiritual nation throughout history to bear precisely the same message and the same essential Divine Energy. This entire venerable lineage, which Islam terms *nur muhammad,* the primordial Muhammadan Light, is the vehicle chosen by the Divine Revealer that manifests as all the Prophets. The third cry of the Call to Prayer contains initiatory power to lift us into the spiritual presence of this prophetic lineage. We are being called to pray side by side with all Divine Messengers, enjoying their sweet spiritual companionship. The Holy Quran refers to this experience as entering the highest circle of the most intimate friends of Allah.

The fourth cry, *hayya-s-salat,* empties our hearts of any limited motivations whatsoever, filling them with the blessed commitment to participate in these infinite praises that come forth spontaneously from the Source of the Universe and flow through us back into the same Source. Responding inwardly to this Divine Call, we leave behind limited personalities, limited projects, limited worlds, just as we remove dusty shoes before entering the place of prayer.

The fifth cry, *hayya-l-falah,* does not refer to the success or realization of any limited aims. It is the invitation from Supreme Reality to enter the mystic state of plenitude, where all powers and attainments are rooted. We are now permeated with joyous expectation. The Divine Gift of *salat* is tangibly descending into the mind, the heart, and even every cell of the body. *Falah,* which literally means *success,* signifies mystic union, the goal of the spiritual path, the ultimate purpose of human existence.

The sixth cry, *allahu akbar,* is the repetition of the first, for the inconceivability of Divine Essence is the central principle of Islam. The seventh and final cry, *la ilaha illallah,* is the key with which to enter Essence.

If one is performing the Call to Prayer in solitude, one can chant intensely in a low voice or cry out like a rush of wind or a peal of thunder. In modern Islamic cities, the Call is transmitted over loudspeakers from the minarets. Whatever the circumstances, Shaykh Muzaffer taught that one's being should open so fully to this Divine Call, whether reciting or simply listening, that were one standing on an iron anvil, it would glow red and melt. This anvil represents conventional self and conventional world.

As Allah reveals in the Book of Reality, were the verses of Quran to descend upon a gigantic mountain, instead of into the diamond heart of His beloved Prophet, the mountain would instantly be reduced to dust. The Call to Prayer is Divine Longing, the very Source of the Quran, descending into the diamond vessel of the secret heart of humanity. If the holy fire of this Divine Longing touched the universe, the universe would be reduced to ashes. Allah Most High directs the Call to Prayer only to the hearts of His precious human beings. It is the promise that once more the Glorious Quran is going to be revealed. *Salat* is that prophetic revelation.

Direction of Prayer

To participate in *salat,* one faces the prehistoric sacred valley of Mecca, associated with the Prophet Abraham. One orients toward the external focus of Divine Presence on the earthly plane, the Kaaba, the cubic structure first built by the Prophet Adam, upon him be peace, swept away and restored several times during prophetic history. Its large blocks of stone are imbued with the fragrance of myrrh, constantly anointed by pilgrims with this precious oil, which the Three Wise Kings brought to honor the infant Jesus, upon him be peace.

For more than half the twenty-three years during which the power of Prophethood manifested visibly through Muhammad the Elect, the twenty-three years during which the noble verses of Quran descended into his heart and then flowed through his lips, the Direction of Prayer remained the holy city of Jerusalem. This indicates how intimately Islam is aligned with its sister traditions, Judaism and Christianity. After the Muslim community retired, by Allah's Permission, from the capitol city of Mecca to the desert oasis of Medina, the change of direction was revealed during the communal *salat* at the moment of the standing bow. The Prophet inclined his blessed head toward Jerusalem the Holy, received revelation, turned while still bowing, and stood up again facing Mecca the Ennobled.

Accurate spatial orientation upon this sanctified planet is an essential aspect of the science of *salat.* Allah Most High, according to Oral Tradition, informed His Prophet that the entire earth has now become a holy mosque, or *masjid,* meaning a place of prostration. Simply to stand facing Mecca the Ennobled unveils the existence of this universal, imageless temple not constructed by human hands. Whether one is at home, at work, in the wilderness or beside a busy street, one prays within this planetary mosque.

Salat is spatial and temporal. The precise periods for prayer, which breathe with the seasons as the times of sunrise and sunset change, orient the faithful by the movement of the sun. The five daily prayers of Islam are sacramental in the sense that they transform human body, earth, and cosmos into vehicles for Divine Blessing, vessels for Divine Energy. *Salat* can never be regarded as merely internal or individual prayer. It is Divine Action flowing through humanity as a whole, in conscious concert with all creation, carefully oriented in planetary space and solar time.

The Direction of Prayer, or *qibla,* is not ultimately a direction within the universe, but the directionless direction of Divine Transcendence. In the planetary circles of prayer about the noble Kaaba, some persons are facing east, others are facing west, some north, others south. The Quran states that Truth is not found in the East or in the West, but that wherever one faces, there is the radiant Countenance of Allah.

Therefore, simply to stand for *salat,* spiritually awakened by the Call to Prayer, gazing into the mystical dimension of transcendence, is already an exalted state of prayer. If there is any difficulty determining the direction of Mecca from wherever one is, one makes the best judgment possible and prays with the firm spiritual intention of facing Mecca. Oral Tradition clarifies that *salat,* offered with this sincere intention, is fully accepted by Allah, even if one may mistakenly be facing the opposite direction from Mecca. This further confirms that the *qibla,* or Direction of Prayer, is not just a geographical orientation but the total opening of one's being to transcendence.

Now one can participate in *salat,* standing with human dignity and uprightness, plunged into a state of inward silence, sober expectation, and liberation from all limited perceptions and motivations. One faces the sacred City of Light, while intending spiritually to experience perfect communion with Divine Light. Nothing can interpose at this moment between the soul and its precious Lord—not the Kaaba in Mecca, nor the wall of the mosque where one is praying, nor the leader of the prayer, nor the lines of the faithful, nor even the physical body or the conditioned mind. The beloved Muhammad reported that not even the radiant angels who surround *salat,* attracted by the beautiful Quranic chanting and the pure intention of the hearts, can come between the soul and its Lord.

The true Kaaba, according to the mystics of Islam, is the secret heart of humanity. When we pray, we are facing the direction of our own innate perfection, the timeless radiance of our preeternal soul. As Allah reveals in His Glorious Quran, humanity is the crown of creation. Allah speaks these words through the lips of His culminating Prophet in a holy *hadith:* "I Who cannot fit into My entire universe can enter easily and completely into the human heart."

Movements of Prayer

By the prayers of Islam, our physical body is transformed into the supreme spiritual instrument. These graceful, rhythmic movements are not simply to accompany *salat* but they are *salat*—the mysterious flow of Divine Energy through which Allah praises Allah and returns to Allah.

The miracle of *salat* is this. It is Divine Power and Mercy become visible and tangible, whether from the perspective of someone within the blessed lines of prayer or even for an outside observer who may know nothing about the tradition. *Salat* is an empirical demonstration of Divine Presence that manifests five times daily in almost all cultures on the planet today. *Salat* is therefore a Divine Gift to humanity as a whole, not just to a particular historical community.

Simply by observing the Movements of Prayer manifesting selflessly through Shaykh Muzaffer Effendi as he performed *salat* in his small bookshop near the Covered Bazaar in Istanbul, many persons were inspired to renew or intensify their practice of *salat,* and some persons even experienced the sudden return of lost faith. This is a spiritual gift called *resurrecting dead hearts.* Members of other sacred traditions, and even open-minded secular persons, have always experienced a mysterious attraction to and kinship with the prayers of Islam. Why? Because *salat* belongs to all humanity.

There are various interpretations for each Movement of Prayer. Oral Tradition teaches that each position represents a particular state of consciousness that the Prophet observed among angelic beings during the radiant night of his ascension. But the transformative power of these primal gestures of praise lies in actually performing them, not interpreting them with the limited intellect or even contemplating them in a mystical context. These movements themselves are rich sources of direct, wordless intuition and waves of ecstatic inspiration, open equally to both the sophisticated and the simple.

One sometimes smells the rose fragrance of the Prophet as one performs *salat,* his unique way of prayer, which is mystically led by him and takes place within his eternal prophetic heart. A single moment of such intimacy with Allah's beloved vastly intensifies the prayerful submission of the personal will to the infinite Divine Will, and is incomparably more fruitful than reading excellent scholarly treatises or exalted mystical poetry.

To lift both hands beside the head, palms open and facing forward, accompanied by the words of Divine Transcendence, *allahu akbar,* is the first distinct movement of *salat.* The open palms signify the openness of the entire being—including every strand of awareness, every cell of the body, and even every pore of the skin—to the Divine Radiance and Magnificence, Who does not fit into the universe yet resides within our secret heart. We are not praying to some limited divinity in front of us, but we are praying within one all-encompassing Consciousness. In fact, it is Supreme Reality praying through us and within us.

According to some masters of *salat,* this first gesture places the entire relative world behind us, including the personality and the will of the one who thinks he or she is performing *salat.* The faithful person is now elevated into soul awareness, surrounded entirely by Divine Light, which is uncreated and undifferentiated. As the Prophet used to supplicate, "May Divine Light be before me, behind me, to right and left, above and below. May my limbs be filled with Divine Light. May my skin be filled with Divine Light."

No description or image of this infinite luminosity can be adequate. Even to say that during *salat* we experience an ocean of light without shores is to impose subtle mental limits. An ocean and one who separately observes it are both created entities. Divine Light is not created, nor is It an entity, nor is It broken into subject and object. This high mystical knowledge is spiritually encoded in raising the hands and crying *allahu akbar.* It is not dependent on verbal instruction and therefore does not become a mere intellectual assertion. This is why *salat* is the most precious gift, chosen for us by our beloved Prophet Muhammad from the Divine Treasury.

The next Movement of Prayer is the union of both hands below the physical heart. This gesture indicates the concentration of the Divine Energy of *salat* at the center of one's manifest being. The right hand is placed over the left, establishing the affirmation of Divine Unity over all possible countertendencies of conditioned awareness.

There follows the chanting of Sura Fatiha, the opening chapter of the Glorious Quran, key to Quran and essence of Quran. According to Oral Tradition, this brief Sura, mystically termed the Mother of the Book, contains the spiritual power of the entire Quran, the cumulative revelation and blessing granted by Allah to 124,000 noble Prophets, beginning with Adam, upon him be peace. The Quran is not simply a revelation that originated with the beloved Messenger of Allah. The Quran contains the entire history of prophecy, and *salat* is the activation of that radiant history in our present experience.

At each of the seven verses, or spiritual steps, of Sura Fatiha, the practitioner of *salat* descends in contemplation through the seven levels of consciousness, from mystic union to separate individual awareness. This follows the way along which Holy Quran descended through the seven heavens into the adamantine heart of the Messenger. This metaphysical route also corresponds to the path of descent of the Prophet Muhammad from the Garden of Essence to bless this planetary sphere, once at birth and again when returning from his Night Journey through the higher planes of Being.

The mystical practitioner of *salat* begins at the seventh station, where Divine Consciousness alone exists, experiencing solely the

Divine Power flowing through the human voice as the first verse, or *ayat,* of the Glorious Quran: *alhamdulillahi rabbi-l-alamin,* all praise flows from and returns to Allah, Who displays countless worlds. The term *ayat* also means Divine Sign or miracle. These seven *ayats* are miraculous signs manifested by Allah as each practitioner chants the Sura Fatiha, silently or aloud, at least thirty-four times a day. On this planet today, there are hundreds of millions of devout practitioners of *salat* chanting these seven verses.

To seal this invocation of the entire Quran, the affirmation *amin* is harmoniously intoned by the congregation, or by individuals when praying in solitude. The *amin* is a distinct act of prayer, the purely human response, *so be it.* Each person must consciously accept this descent of Divine Energy that constitutes *salat,* because the drama of human and Divine is a synergy.

At this point the process of prayer reaches a high level of intensity. The precise Message and Blessing of Allah descends as certain verses from some six thousand verses of Quran. The selection is limited by the number of *ayats* known by heart, either by the individual praying in solitude or by the leader of congregational prayer. If the Imam is a *hafiz,* one who knows the entire Quran by memory in a manner as natural and easy as we remember our own name and place of birth, the selection is unlimited.

One *hafiz* reports that as he is completing the musical arabesques in the final verse of Sura Fatiha, the sound becomes like a bee, slowly descending upon a particular Quranic verse, as upon a flower of light in a vast Quranic rose garden. The actual *ayat* is revealed to his mind only at the moment of the *amin.*

Even if our selection of Quranic verses is extremely limited, we must recognize Divine Initiative in revealing the *ayat* at the moment of *amin.* We must cultivate the sense that these are Allah's Words we are chanting, Arabic sounds that bear Divine Energy, not human words filled with our own energy. Not only the various levels of meaning but the very resonance of the Arabic Quran is the Divine Resonance, streaming through the cooperating human instrument.

The next Movement of Prayer is the standing bow, hands resting on knees. The communication between nerves and muscles by which this deep bow is accomplished should be experienced as coming from Divine Power. Human intention proceeds from Divine Permission. As one pauses in the fullness of this bow, immersed in great reverence for Allah, certain words of praise first used by the beloved Prophet are whispered softly, with intensity and intimacy, as lovers whisper to one another. Allah Most Near manifests as absolutely near to the one who bows in *salat*—nearer than the hearts of lovers who experience emotional separation as well as affection, nearer even than we are to ourselves. This exclamation arises from joyful submission of one's will to Allah Most High, rather than expressing any individual supplication, which is part of personal devotion, not *salat*. Three times the prescribed phrase is repeated: *subhana rabbi-l-azim,* glory to the Supreme Teacher Who is sheer magnificence.

With the unhurried rhythm characteristic of *salat,* Divine Power again lifts the human instrument to the standing position, the sublime upright bearing of the true human being, reflecting the innate goodness and eternal dignity of the soul. While rising, one intones a prescribed cry of praise and expresses a further affirmation of gratitude when reaching the upright position, using a venerable Arabic phrase that first came from the lips of the noble Ali.

The next Movement of Prayer, punctuated as the others are by the cry of transcendence, *allahu akbar,* is an expression of total trust and total abandon. One releases any trace of separate individuality and plunges, with delicacy and balance, into the immeasurable ocean of full prostration. As the forehead makes contact with the earth—whether wooden or marble floor, prayer carpet, or simply bare ground—the soul melts into its Original Source. Another divinely revealed form of praise is whispered slowly, without any sense of personal initiative or even individual existence. Three times the words repeat themselves, without subject or object: *subhana rabbi-l-ala,* glorious is Supreme Reality.

Now Divine Power miraculously draws one forth from the ocean of union into a steady seated position, expressing the mystical state of diamond permanence within Allah. The human instrument is kept balanced there, like a pendulum in slow motion at the edge of its arc, and then is returned once more into complete prostration. Only the power of Allah could accomplish these movements, because there is no sensation of limited will. There is no ability or desire, only subtle rapture.

With the cry of *allahu akbar,* one is lifted miraculously from full prostration to the standing position. It is like the Resurrection. One stands without losing the sense of full prostration, the actual experience of submission to Allah, which is Islam. This cyclical process now repeats again, including the Sura Fatiha, the *amin,* and the other self-revealing *ayats* of Quran. But the touch of prostration remains on the forehead, where Divine Light is now streaming forth, for those whose heart's eyes are open. As the Messenger of Allah remarked most simply, "Prayer is Divine Light."

After the second cycle of prostrations, one remains seated on the heels, like a pendulum miraculously suspended at the edge of its arc. This is not a position of rest but blessed concentration and elevation. In this dimension of diamond clarity within Allah, one greets face to face the Prophet Muhammad, upon him be peace, the Prophet Abraham, and the entire prophetic lineage, including the circles of their loyal and fervent companions throughout history. But most intimately one faces the beloved one of Allah, greeting him in an Arabic grammatical form suitable only to close proximity: *as-salamu alayka ayyuha-n-nabi,* peace be upon you, O sublime Prophet.

Alive within Allah by Divine Life alone, one now witnesses Supreme Reality through the infinite *I Am,* no longer through the reflected *I am.* Even the physical position remains one of witnessing, as the right index finger and the big toe on the right foot both point in the mystic Direction of Prayer.

The prescribed words for greeting the Messenger of Allah, for witnessing Divine Unity, and for calling Divine Blessings upon all the noble Prophets and their spiritual families are repeated in Arabic, whether one is praying in Africa, Arabia, Russia, India, China, Indonesia, Europe, or America. The effect of using the ancient Arabic is to remove one from the personal associations of native language and to lift one into universal praise, unveiling the soul as a timeless companion of the Messenger of Allah, greeting him intimately in his own Arabic idiom. The masters of *salat* consider these Arabic greetings to be a conversation between Allah Most High and His beloved Prophet during his ascension, rather than a personal expression by the one who prays.

By participating in the original Arabic form of *salat* precisely as it was revealed to the Prophet of Islam, one is lifted above the play of cultural and intellectual preferences. One is also lifted above the powerful current of one's personal imperatives—biological, social, and even religious. *Salat* can therefore be said to transcend culture and religion, in the sense that they are mere human institutions, mere collections of inhibitions and obsessions. This is why *salat* travels so easily among historical societies, including the Eurocentric culture of the modern world, while continuing to retain its basic form, integrity, and effectiveness. *Salat* frees the human spirit from self-imposed limits.

Manifesting Divine Power of transformation and elevation, *salat* achieves the sanctification of space and time, the opening of the mystic path to everyone without distinction, and the unveiling of true humanity. Allah alone is capable of this miraculous feat, whether in the ancient world or in the modern and postmodern worlds. The original life of prayer, revealed directly to the beloved Muhammad and practiced so carefully by him, has been transmitted through bodies and minds of fourteen centuries of Muslim practitioners, and is now intertwined with our entire planetary civilization.

There exist at least four major schools of Islamic traditions concerning the performance of *salat,* each one recognized by the others to be genuine, that is, rooted in the life experience of the Prophet and transmitted through an authentic line of his spiritual successors. The beloved Muhammad, upon him be peace, lived in constant ecstatic communion and union with Allah Most High. He prayed in many different manners. Therefore one finds healthy and refreshing diversity as well as fundamental agreement among these schools of Islamic jurisprudence. *Salat* always manifests its unique spirit, which can be recognized instantly in whatever Islamic cultural environment one enters, regardless of variations.

The promise of *salat*—to establish the direct experience of Transcendence in the hearts and daily lives of all people, without discrimination on grounds of gender, race, or social status—has remained good throughout fourteen centuries of complex Islamic history, filled with the struggles of personal and collective egocentricity. Human institutions and human individuals have failed, never *salat*. The Divine Promise of *salat* is good until the End of Time.

The final gesture of *salat* is the Word of Peace repeated twice, first to the right along the lines of prayer and then to the left: *as-salam alaykum wa rahmatullahi,* may the sublime Peace and compassionate Blessings of Allah be with you. The Peace and Mercy that have descended directly from the Garden of Essence now flow outward, through this final Movement of Prayer, to everyone in the vast planetary circles of *salat,* to angels who gather to witness and enjoy the miracle of *salat,* to every human being on the face of the earth, to subtle beings on other planes of existence, and to every creation of Allah. *Salat* is the gift of Peace and Mercy to all consciousness.

Through this culminating gesture, Allah transforms the participant in *salat* into a perfect instrument of Divine Peace, consciously immersed in the peace that comes from directly experiencing Supreme Reality. The Arabic word for peace, *salam,* indicates the essence of Islam, which is not a religious organization but an organic experience of entering Divine Peace through the open door of *salat.* The responsibility of Muslim men and women, whose daily lives have been thoroughly transformed by the dignity and radiance of these Movements of Prayer, is to share their wonderful experience of peace with all conscious beings, not verbally but existentially.

The highest meaning and function of religion is to be an open door through which the soul returns consciously into Divine Essence, not simply after death but already during this earthly life, which does not exist outside Divine Life. Authentic religion, therefore, always transcends its limited external forms by means of its own sacramental power, its own openness to limitless Reality. However, the organic form of the central sacrament—in the case of Islam, the noble *salat,* the open door into Divine Essence—cannot be tampered with or dispensed with, but must be wisely protected and sustained.

Islam is to experience Divine Peace through the daily practice of prayer and prayerful living. This is the base of all religion. Thus, following the Holy Quran, we can speak of the universal Islam that abides at the heart of all the noble traditions revealed during human history. In this universal sense, Islam is the very design of the human soul, the secret of fruitful civilization, the inner functioning of creation. Never does the Quran suggest that Islam originated with the beloved Muhammad of Arabia. What is unveiled by the Prophet is *salat,* the active principle and transparent expression of the dynamic and essential peace called Islam.

Now *salat* is complete. The peace of conscious communion with Supreme Reality is transmitted once more to all creation. While still seated, the practitioner may recite various Quranic passages and may offer personal supplications. These forms vary slightly from one Islamic culture to the next. However, with the final word of peace, the fullness of *salat,* the miraculous descent of Divine Peace and Mercy, has once again been accomplished by Allah Most High.

Standing and reentering the multidirectional world of relativity, heart still facing the Direction of Prayer, one gives the kiss of peace or extends the hand of peace to all those who have shared this blissful communion with Divine Peace. One is now irradiated by inexpressible peace. Although invisible to ordinary eyes, the Light of Islam streams forth from the foreheads of those who have truly prostrated.

Having been imbued by the graceful rhythm of the prayers, one never rushes to *salat* or rushes away from *salat.* The Prophet, upon him be peace, once warned a Muslim to slow down his prostrations because he resembled a chicken pecking grain more than a true human being communing with the Source of the Universe.

One comes forth from the place of prayer to engage fully in life—as the Quran states, to experience the abundance of Allah. Yet one subtly retains the state of mystic union and instinctively anticipates the next opportunity to pass through the mysterious portal of *salat,* a precious opportunity that will occur again not in a year, month, week, or day, but within a few hours.

Times of Prayer

All creation is immersed in ecstatic praise of its Creator. The five daily prayers of Islam establish conscious connection with this cosmic dimension. The quality of light at each divinely ordained *salat* invokes a particular state of consciousness and hints at certain spiritual gifts that this period of prayer contains. The times move a minute or two each day as the sun subtly changes its orientation to the earth. *Salat* is a spiritual astrolabe, a way for the timeless soul to navigate successfully through the temporal cosmos. Those who participate in *salat* experience the personal and cosmic harmony that flows mercifully from the Source of Being.

We will review the basic *salat,* not taking account of various categories of supplemental *salat.* We are presenting here merely an atom from the sun of knowledge.

Dawn Prayer *(fajr)*
two *raqats,* or cycles of prostration, prayed aloud

This period for *salat* arrives not during predawn or false dawn but when the first white light extends across the horizon. Certain species of birds know as accurately as astronomers the first moment of *fajr.* This blessed period of prayer extends until just before the sun breaks the horizon. The compassionate Oral Tradition, focused by Allah Most High through the mind and heart of His culminating Messenger, reveals that dawn prayer may be accepted by Allah if offered sincerely whenever one awakens, as long as it is before the time for noon prayer. The spiritual quality of *fajr* is purity of intention, as one is newly born into the Divine Light we call creation.

Noon Prayer *(zuhr)*
four *raqats,* prayed in silence

This powerful period for *salat* arrives when the sun reaches its zenith, crowning and sanctifying the intensity of the morning's work in the world. This prayer transmits the sense of Allah's Sovereign Power and Divine Clarity, as well as unveiling human dignity, rectitude, and plenitude. The mysterious door of the noon *salat* remains open until the time for afternoon prayer. It is prayed in silence to intensify inwardness.

Afternoon Prayer *(asr)*
four *raqats,* prayed in silence

This central prayer period of the day opens when the sun's rays slant diagonally, indicating the beginning of its swift descent toward the horizon. The quality of light and birdsong changes distinctly. At this point, one is in the greatest danger of becoming absorbed in mundane activity, of losing one's longing for *salat,* for constant remembrance of Supreme Reality. The Prophet, upon him be peace, therefore proclaimed, "Guard well the middle prayer." The quality of this *salat* is the refreshing energy of transcendence. It is prayed in silence to intensify inwardness.

Sunset Prayer *(maghrib)*
two *raqats* aloud and one in silence

The period for sunset prayer is brief, like that of dawn prayer, but without the dispensation to complete it later. Therefore, there is a special concern to perform this *salat* immediately when Allah Most High opens the door between dimensions with the key of the Call to Prayer. The Call is given after the sun's orb disappears below the horizon, with adjustments made for different configurations of the earth. Like dawn prayer, sunset prayer unveils the light of Paradise, an intrinsically radiant landscape where there need be no separate source of light, where there is no appearance of any sun. Ease, sweetness, fragrance, coolness, serenity, and bliss, the characteristics of Paradise, are experienced during both dawn and sunset prayers, but especially *maghrib.* Both sets of prayer are chanted aloud, for the breeze in Paradise blows through perfumed trees with the sound of Quranic verses.

Dawn prayer comes without effort, except for the discipline of arising early, whereas sunset prayer arrives after a long day of mundane efforts and spiritual struggle to maintain inward remembrance of Allah. *Maghrib* is experienced as a liberation, an immense happiness, a spiritual victory. Thus it replicates more exactly than *fajr* the experience of entering Paradise. The serenity and completion radiated by the sunset prayer heals the entire heart and mind, just as twilight coolness refreshes the body after the heat of the day. One's entire being is filled with the youthfulness of Paradise consciousness.

The noble lines of prayer always gaze toward Paradise, but the sunset prayer occurs at the very threshold of this most sublime state of awareness. However, *salat* remains firmly planted on the earth. Within the transcendental precincts of Paradise, there is no cosmic temporality, no movement of any sun, and hence no *salat.* Within Paradise consciousness, which can be glimpsed even while living on earth, every action and perception is direct Divine Communion.

Shaykh Muzaffer Effendi used to remark that the only disappointment for the faithful person upon passing away from this planetary sphere is the necessity to bid farewell to the glorious and beloved *salat.* All other loved ones will be encountered again, for the Prophet proclaimed, "You will be with the ones you love."

From the perspective of eternity, we are presented only a flash of temporality within which to enjoy *salat,* to participate in its graceful power and heart-melting beauty. It is for this reason that the prayers of Islam are performed by knowers of Truth not as a tiresome obligation or a protection against hellfire, but as the joyous play shared by lovers. Even if persons do not always participate in formal prayers, *salat* is still going on in the profundity of their being and in the depth of human being. Humanity has become *salat.* This is why Muhammad, upon him be peace, is called the Seal of Prophecy. There is nothing more for

Allah to reveal, nothing more for humanity to realize. We are living together in the fullness of time.

The brief period for the sunset prayer, which expands during summer months, extends until the last and most subtle drop of twilight has disappeared into the rich, radiant blackness of night.

Night Prayer *(isha)*
two *raqats* aloud and two in silence

This is the *salat* of Divine Mystery, which leads beyond the far boundary of Paradise into the Garden of Essence, the sublime annihilation of limited existence. Sufi masters call this the *black light*, symbolized by the black cloth that covers the Kaaba and by the black stone enshrined in silver at one corner of the Kaaba. This imageless, uncharacterizable blackness is beyond Divine Light, which is an Attribute of Allah, not the Essence of Allah. Divine Essence is impenetrable by human intellect, even when guided and illumined by revelation. And this rich, radiant darkness is beyond the reach of mystical experience. Angels and even Divine Attributes face and reflect Essence without entering there. Only the human soul, which is the created expression of Divine Essence, can abide consciously within the mysterious Garden of Essence. This entering into Essence was demonstrated and activated for all humankind by the Seal of Prophecy during his mystical Night Journey. Into the mystic darkness of the *isha* prayer, the original Divine Mystery, the ray of our separate human awareness disappears.

The period for night prayer extends to an hour before first light, leaving the door to Supreme Identity wide open all night. The noble Prophet, upon him be peace, preferred to make *isha* late, without sleeping beforehand, explaining that simply waiting for these unique night prayers is an exalted state of worship. This is the only *salat* he would regularly postpone until well after the prayer period arrived, indicating the special power and importance of *isha*.

The person who lives in the dynamic current of these five daily prayers exists in a transfigured temporality, upon an earth that is like a rich prayer carpet spread out for all humanity, precisely as the Holy Quran describes this blessed planet. Spatial direction and the sun's movement have become potent reminders of prayer. In fact, they have become part of prayer. Sun time is unveiled as prayer time, rather than remaining a mere biological clock marking the conventional hours for eating, working, sleeping, and waking. The whole earth has actually become a vast, sacred mosque, as Allah informs humanity through the Oral Tradition of His Messenger. The entire creation has been revealed as pure consciousness, as pure prayer.

Ablutions before Prayer

In the transfigured earthly existence created by the intense atmosphere and delicate fragrance of *salat,* water takes on the sacramental power of purification and transformation. Because *salat* is the ascension of the faithful, it transports lovers to the transcendental palace of the Sultan of Love. Before entering the intimate presence of the cherished Sovereign, we are rightly concerned to prepare ourselves. *Salat* is meeting the Holy King, not in the spacious audience hall where every conscious creature can come, but in the private royal chambers, opened for all humanity by the beloved one of Allah.

How fresh, how pure, how unencumbered we must be to enter there! There is no earthly garment magnificent enough for this meeting, this communion, this union. We not only remove our sandals before approaching the burning bush of *salat,* but our entire body and mind must be transformed. Once we really know where we are standing, we can no longer wear the stained robes of conventional consciousness. We would be shocked to observe someone in the mosque wearing muddy boots and making the movements of prayer. But these boots, covered with soil from the sanctified earth, are much more pure than an egocentric awareness that prays to be seen by others, to maintain cultural identity, or to escape hellfire.

How can we obtain a robe of stainless, selfless awareness in which to perform true *salat?* Allah alone can provide this, as He provides every miraculous gift by which we exist and praise. The ablution before prayer is the royal wedding garment with which we can authentically enter Divine Presence and delight in the communion desired by the mystical King.

The ablution of Islam is not a form of washing but a transformation of the whole being, including physical body, etheric body, emotional body, mental body, and subtle body. This ritual purification is not merely symbolic but is accomplished by water that has become waves of Divine Blessing, or *baraka*. That the spiritual state of ablution does not come about through washing with physical water is further confirmed by another Divine Permission: the valid sacrament of ablution may be performed with clean dust from the earth whenever water is not available.

An Oral Tradition of the Prophet, upon him be peace, reveals that the spiritual body in Paradise, which is composed purely of Divine Light, manifests special brilliance at those places—such as face, forearms, and feet—that have been touched repeatedly during temporal existence by the empowered water of ablution. Our own experience directly confirms this teaching concerning the transformation of the earthly being by ablutions. One feels total refreshment and renewal of body and spirit after performing the ablutions of Islam in the name of Allah Most High, regardless of the mundane pressures and distractions that constantly impinge on consciousness. This can be tested and demonstrated again and again.

As well as preparing us for the mystical ascension of *salat,* the sacrament of ablution can be used to counteract anger or depression. Persons on the steeply ascending path to union with Truth renew their ablutions perpetually, not only to prepare for *salat* but to perform the prayerful act of human existence itself. Sufi saints renew their ablutions if they forget Supreme Reality even for an instant or become even slightly distracted by the surface world.

The ablutions before prayer need to be renewed after acts of elimination and passion. This does not imply any negative judgment that the natural world is impure. All the creations of Allah are essentially clean and pristine. Beneath the opaque surface of the conventional cultural landscape that we usually perceive lies the perfect Creation of Allah, composed solely from Divine Attributes, manifesting harmoniously through inconceivable Divine Wisdom and Mercy. The noble ablutions of Islam permit us to enter consciously into this radiant Garden of Creation. As Allah Most High instructed His beloved Prophet in the Glorious Quran, "Send forth your gaze into My Creation. Can you encounter the slightest imperfection, injustice, incoherence, or disharmony there?" This Quranic passage goes on to reveal that the powerful gaze of the Prince of Prophets returned from this journey through the depth of Divine Creation astonished and overwhelmed.

We renew ablution after sleeping because the ablutions of Islam create perfect wakefulness to the Divine Call. We renew ablution after shedding blood because the ablutions of Islam constitute the healing and protection of our earthly being. These ablutions help us to stop identifying ourselves as limited bodies and conditioned minds. This clarification by water allows us to perceive, with eyes of the heart, that we are limitless souls, incarnated in the beautiful environment of Divine Manifestation. Only with this purified perception can we truly participate in *salat.*

The ablution is as essential to *salat* as standing to face the noble city of Mecca. Just as the spatial direction of prayer, while remaining geographical fact, is transformed by Allah into the mystic direction of Divine Transcendence, so the water of ablution, while remaining the precious earthly substance that quenches our thirst, is transubstantiated into the purifying and transforming power of Divine Light.

Dua, Salat, and Dhikr

Three dimensions of prayerfulness exist in Islam. *Dua* is the concrete personal prayer of petition for oneself, for family and friends, for all humanity, for all conscious beings. After the completion of *salat,* there is an open space for *dua,* but *salat* itself maintains neither a personal nor a petitionary focus. *Salat* is pure praise, affirmation of unity, mystic communion. *Salat* is not our subjective, individual form of supplication or gratitude but rather the objective, communal science of ascension and union. *Salat* is more a vast, cosmic laboratory of Divine Demonstration than it is a private, inward devotion.

Dua, or supplication, can seek with confidence from the All-Merciful the fulfillment of any longing and can be resorted to at any time, with or without the performance of ablution. In addition, what may be termed *prayerful speech* pervades the entire daily life of Islam in the form of various characteristic affirmations: *insha'allah* (Allah willing), *masha'allah* (may it be pleasing to Allah), *alhamdulillah* (all praises are due only to Allah), *estaghfirullah* (may Allah forgive us), and many others. Even though Allah is the most holy Divine Name in Islamic tradition, the people of *salat* are encouraged to intersperse and beautify their speech with this powerful utterance. Such traditional phrases can be used merely conventionally by the immature or with great sensitivity by those who are spiritually mature. In either case, the beautiful Divine Name Allah carries transforming power. Even the most mundane spheres of communication in Islamic culture thus become filled with the atmosphere of prayerful speech.

The third dimension of prayerfulness is *dhikr,* which simply means remembering. With grateful love and awe, the spiritually awakened person constantly remembers the existence of Allah Most High, remaining in this blessed state of recollection or recollectedness during every form of activity, even while dreaming. Clearly, it is *salat* that makes this profound realization possible, not just for a few highly trained contemplatives but for millions of devout Muslims in every generation of global Islam.

Dhikr is a unique way of remembering—not like remembering the time for prayer or recalling other important information, but more like remembering that one exists. In the spiritual station of *dhikr,* this primordial sense of consciously existing is focused on Supreme Reality, which is the Only Consciousness, manifesting as innumerable conscious beings. Sufis call this the Eye of Certainty. In the true state of *dhikr,* there cannot exist even a thread of doubt or an atom of separation. Interpreting the noble Quranic verse "If you remember Me, I will remember you," the mystic recognizes *dhikr* as the Divine Remembrance, as Allah remembering—throughout time, as well as timelessly. *Dhikr* resembles *salat* in the sense that it belongs primarily to Allah, not to humanity. Our individual personhood is expressed through *dua.*

Unlike *salat* but like *dua, dhikr* is appropriate in every moment, under every condition, at least inwardly. But *dua* is a state that comes and goes, whereas *dhikr,* when mature, becomes a permanent spiritual station. *Dhikr* may manifest at various times, with various levels of intensity, either externally or inwardly, formally or informally, but it remains essentially continuous, indivisible, without subject or object.

Salat, even when it enters mystic union, retains a strong characteristic of praise, whereas *dhikr,* although operating through the repetition of various Divine Names, "leaves the names and comes to the One Who is Named," as our Grand Shaykh Muzaffer Effendi used to say. In *salat,* creation is praising its Creator. In *dhikr,* creation is unveiled as a tapestry of Divine Names, or essential energies, which are constantly turning toward Essence, of which they are emanations. In *dhikr,* the veils of Creator and creation fall away.

Ablutions are not technically necessary for the performance of *dhikr,* which belongs to the secret heart, always elevated and pristine. However, if one is performing *dhikr* in congregation, joining a circle of dervishes under the guidance of a shaykh, one establishes ablutions out of respect for the mystic guide. One would probably already have ablutions, in fact, since the traditional ceremony of Divine Remembrance is usually conducted just before or after night prayer. However, Shaykh Muzaffer Effendi, when operating in secular western society, would invite the public into his circle of *dhikr* without even requesting them to remove their shoes. This confirms the essential freedom of *dhikr* from any restrictions or ritual context whatsoever.

Dhikr may also be practiced silently or aloud in solitude—while sitting, walking, standing, or lying down—either following the guidance of a shaykh, who is a living link in the mystical succession that extends back to the noble Ali, or simply by the inspiration that comes to every soul from Allah Most High. The practitioner of *dhikr* may use the prayer beads characteristic of Islam or simply the living beads of each breath or heartbeat. *Dhikr* can be continued for a specific number of repetitions of various Divine Names, or it can extend continuously, awake and asleep. The central form of *dhikr* is *la ilaha illallah,* nothing exists apart from Allah.

Salat, according to prophetic custom, is followed by a brief communal *dhikr,* usually accomplished in silence. Counting on beads or with fingers, the congregation or the solitary person repeats thirty-three times each the fundamental affirmations of Divine Magnificence: *subhanallah, alhamdulillah, allahu akbar.* Although this particular practice descends from the Prophet of Islam and contains many secret spiritual benefits for the individual practitioner and for humanity as a whole, it is not a formal part of *salat* and therefore remains optional.

Following the reliable words of the beloved one of Allah, "*salat* is the greatest *dhikr,*" we can conclude that no matter how fervent one's supplication or how intense one's Divine Remembrance, *salat* remains the base and heart of Islamic spirituality while on this earth. In Paradise, *dhikr* becomes central, for there can be no more *salat,* only direct Divine Presence. In the Garden of Essence, beyond the far boundary of Paradise, even *dhikr* ceases.

When we participate in *dhikr*—communally or individually repeating *la ilaha illallah, allah,* or simply *hu*—we are losing ourselves in timelessness, abandoning our relation to temporal existence, which is the praise of the sublime Creator by the humble creature. By contrast, *salat* is always rooted in time, in creation, in praise. By *salat,* temporal existence is stripped of its conventional, egocentric veils and becomes the blessed experience of Divine Holiness, Divine Sovereignty, Divine Mercy pouring through creation, from the far horizons of the universe to our own intimate hearts. Although sanctifying temporality, *salat* faces eternity.

Salat establishes five royal residences in the landscape of each day: *fajr, zuhr, asr, maghrib, isha.* At each of these resplendent palaces, we are offered intimate audience with the Supreme Sovereign of all realms, regions, and dimensions. What would be a memorable event even once in an entire lifetime, meeting with the mystical King, becomes an awesome and delightful opportunity five times every day. May we appreciate even an atom of the honor and sublimity conferred upon humanity by Allah Most High through His noble Messenger!

MEVLUD The *mevlud* is a literary genre that appears in local languages throughout the vast expanse of global Islam. The most sublime poetry, musical performance, and purified emotion of Muslim tradition gather around this expression of devotion for the Prophet and his family. The particular version interpreted here into an enriched English was composed by the Turkish mystic saint Suleyman Chelebi (d. 1419). Including the miraculous signs at the natural birth of the Prophet and his mysterious ascension through the various planes of Being into the unimaginable Source of Being, this poem is chanted on such illumined occasions as the Prophet's Birthday and his Night Journey, or *miraj*. This poem is sung or recited as well to celebrate the birth of a child or the passing of a soul. The present English version was first read aloud by its author during the early spring of 1985 at the Mosque of Delight (the Masjid al-Farah in New York City) to celebrate the fortieth day of Shaykh Muzaffer's passing from this reflected world into the original realm of Divine Beauty.

3
Mevlud
Mystical Biography of the Prophet

Way of Truth and Light of Prophecy

> *bismillah ir-rahman ir-rahim*
> May every action, every thought, and every breath
> be taken in the most holy Name of Allah,
> Who is infinitely tender Mercy and boundless Compassion.

We sincerely invoke the Name of Majesty, Allah Most High. This beautiful Name is always on the lips of the servants of Allah. This all-powerful Divine Name is given even to the most humble, bearing the sweetness of the All-Merciful and the invincible strength of the All-Sublime.

Were every act and intention taken with *bismillah ir-rahman ir-rahim,* humanity would immediately reach perfection.

May we breathe *allah allah* with every breath, bringing the entire universe to spiritual fulfillment. The servant who repeats the Name of Allah with love will see all sense of sin and separation swept away like leaves in autumn. The Name of Allah is most pure and instantly creates purity of heart in the one who breathes *allah allah* unceasingly, thus fulfilling the highest purpose of human life.

Come, come to the rapture of Divine Love! Come brothers and sisters, shedding tears of love and overwhelmed with awe, so that the Sultan of Love, supremely Generous and inconceivably Holy, may shower His sweet Mercy upon us all.

Allah is perfect Unity. Never allow the spontaneous affirmation of Unity to disappear from your awareness, no matter how many deluded teachings proclaim disunity. Before all worlds and before eternity, Allah alone is. This Divine Completeness is incomparably more complete than creation with its human beings and angels, earth, moon, sun, subtle realms, and highest Paradise. Creation flows only from Allah, as the revelation of Divine Power and the demonstration of Divine Mercy.

At the simple Word of Power, *Be!* the entire universe shines forth instantly. Were Allah to will, the same universe would vanish again without a trace. The entire spectrum of creation exists only because of the beloved First Light, *nur muhammad,* may peace eternal be upon him. With open hearts we turn to the holy Prophet of Allah for illumination.

To be drawn into the Light of Allah and not be blinded, O brothers and sisters, repeat with longing and with love:

> *as-salatu wa-s-salamu alayka ya rasulallah*
> *as-salatu wa-s-salamu alayka ya habiballah*
> Heartfelt and profound greetings to your sublime soul,
> O incomparable Prophet and uniquely beloved one of Allah.

Welcome, worthy friends of Allah, gathered here only out of love. May Allah Most Sublime fill with ineffable fragrance the souls of those who pray for me, your slave in mystic love and the composer of this hymn. Please offer together now the opening Sura of Quran, the noble Fatiha, master key that opens the treasure of Divine Love.

Allah brought forth humanity as the brilliant crown of His creation. Endless ranks of angels bowed before Adam at the Divine Command. On Adam's forehead shone the Light of Guidance that descended from the preeternal Muhammad of Light. This original Light of Prophecy has been transmitted by the noble Prophets of all nations throughout history, culminating as the beloved Muhammad of Arabia.

The Light of Prophecy remained unwavering on Prophet Adam's brow until he departed from this world, and then this Light moved to the gentle brow of our holy mother Eve. Next the Prophet Seth, upon him be peace, received the primordial

Muhammadan Light, which increased its manifest radiance as the generations passed. Thus the holy Prophets Abraham and Ishmael, upon them be peace, received this precious Light of Revelation. The lineage of this Light extended from brow to brow in unbroken transmission through 124,000 Messengers, including Moses and Jesus, upon them be peace, until it reached its culmination as the Pearl of the Universe, the temporal Muhammad, Mercy of Allah to all Worlds.

Such is the transmission of Divine Light from preeternal Light to historical culmination of Light. Be in awe of the resplendence of the Prophet of Allah, O lovers, and be confident in the power of his compassionate intercession.

Descent of the Prophet's Soul

Blessed Amina, tender mother of the Pearl Beyond Price, is with child by Abdullah, faithful servant of Allah. The hour of nativity is approaching. On this mystic night when the soul of the holy Prophet descends, there are wonderful signs of the Good Pleasure of Allah Most High.

Amina recounts, "I see beautiful light surging from within me and streaming from my house with ever-increasing intensity. The sun circles around this Divine Light like a dazzled moth. Earth and heavens shine with gemlike clarity and splendor.

"The radiant Gates of Paradise open wide, and the principle of darkness is overcome. Three angelic beings appear, bringing banners of light, which they raise in triumph in the East, in the West, and upon the holy Kaaba.

"Countless ranks of angels descend and circumambulate my humble dwelling as though it were the sacred Abode of Allah. This angelic multitude reveals to me that their Leader and Shaykh will now manifest on earth to bless humanity. In this mode of spiritual vision, I see a silken couch descending, attended by angels lost in adoration.

"My heart overflows with joyous wonder. Through the wall of the room step three shining *houris,* wisdom maidens of Paradise.

The first is blessed Asiya, who nurtured the Prophet Moses. The second is the sublime Lady Mariam, Virgin Mother of the Messiah Jesus. The third is their graceful attendant. With foreheads clear as the full moon, they approach me and, bowing low, utter in sweet and loving tones, *as-salam alaykum,* may Divine Peace be upon you. They sit joyfully around me, while each announces the coming and sings the praises of Muhammad Mustafa, the Chosen One."

"They say to me: 'Never since the creation of the world has a mother had such cause for exultation. No son like this son of yours, with such fullness of power and grace, has ever been sent by Allah Most High for the healing and illumination of the earth and its peoples. You have found supreme favor with Allah, O delicate and lovely Amina, for your son is a spiritual sultan, possessing the secret knowledge of Truth and perfectly expressing in his very being the affirmation of Divine Unity. Humankind and angels are longing to gaze upon his luminous countenance. Even the heavens are turning toward him in awe as his shining soul descends. This is the supreme night promised by Allah and the Prophets of Allah. The whole creation is rejoicing, for manifest Being is miraculously transformed tonight into Paradise to receive this glorious soul, the Mercy of Allah. This is the night that mystic lovers have long awaited. All persons of love are so elated they cannot sleep. Muhammad Mustafa is the sanctuary for all who are lost and wandering. He is the tender Divine Compassion that unveils Paradise even in this world.'

"Thus the circle of holy women speak to my heart, setting it on fire with love and yearning. I thirst. I thirst for Truth. I burn with the fever of ecstasy."

A brimming cup is brought at once to blessed Amina, a drink of perfect clarity, the snow-fresh coolness of perfect wisdom. No nectar was ever so sweet and satisfying.

"Gratefully, I drink this water from the central spring of Paradise. Divine Light fills my entire being. I cannot continue to distinguish my separate self from this all-encompassing Radiance."

Rise in spirit, brothers and sisters, for at this moment our mother Amina perceives a white dove, descending on brilliant wings that stroke her back with delicate strength. The Sultan of Faith is given to humanity at this very hour. Earthly plane and heavenly realms are drenched in a luminous rain of Divine Grace.

Turn to him, approach him, offer him your salams and your complete dedication. Paradise will spring from your tears of longing. To be drawn into the Light of Allah and not be blinded, O brothers and sisters, repeat with longing and with love:

> *as-salatu wa-s-salamu alayka ya rasulallah*
> *as-salatu wa-s-salamu alayka ya habiballah*
> Heartfelt and profound greetings to your sublime soul,
> O incomparable Prophet and uniquely beloved one of Allah.

All creatures now sing greetings to the Prince of Prophets. The sorrow of the world is dissolved and the universe flooded with new life. The mountains and the stars join the circle of conscious beings crying out with sweet, melodious voices: *marhaba ya rasulallah*. Welcome, Prophet of Allah. Welcome, incomparable Sultan of Prophets. Welcome, Fountain of Mystic Knowledge. Welcome, Treasury of the Secret Meanings of the Holy Quran. Welcome, Healing Balm for all the world's afflictions. Welcome, Nightingale from the Garden of Divine Beauty. Welcome, Bearer of the universal Forgiveness of Allah.

Welcome, Moon and Sun of Allah's Mercy. Welcome, Perfect Focus of the Truth. Welcome, Shelter for the wandering soul. Welcome, Unwavering Spirit. Welcome, *ya saqi,* Cupbearer of the Wine of Love. Welcome, sublime Vision of those who truly adore. Welcome, Prince of Humankind, Beloved of the Creator. Welcome, Illumination of longing souls. Welcome, most intimately cherished Friend of Allah. Welcome, most beloved Soul. Welcome, Intercessor for the entire universe.

Welcome, O Sultan of Prophets for whom the earth and Paradise were both created. Your face shines with the intensity of noonday sun. Your powerful right hand lifts up those who have fallen along the way of Truth. You are the only foundation and solace for the lovers of Reality. You are the spiritual refuge for all conscious beings. You are healing for hearts broken by the world. You are the delicate spiritual joy of the heart that turns away from the world. O Sultan and Shaykh of all creation. O President of the Parliament of Prophets and Mystic Guides. O Clarifying Light for the eyes of human beings, both saints and simple believers. O Final One to mount the Throne of Prophecy. O Seal of the Prophets and Distributor of the Light of Prophecy to all hearts.

Your face makes the whole creation shine with greater clarity and brilliance. Your roselike beauty fills the hearts of lovers with the fragrance of roses. You overcome the night of spiritual ignorance and error, bringing the vineyard of realization to fruition with the sunlight of the Holy Quran. O beloved one of Allah, please assist us along the mystic way, and permit us to gaze upon your Face of Beauty at the End of Time.

When the Messenger of Allah, the perfect reflection of Divine Mercy and Compassion, shines on earth with incredible beauty, each angel sings out in delight, and the very earth trembles in sweet transport. Blessed Amina is lost in awe and amazement. Entering total ecstasy, she is granted a vision. She finds herself suddenly alone in the transparent presence of Allah. The wisdom maidens have vanished and with them her newborn son. The sweet mother feels no attraction for this state of mystic union and prays fervently to Allah to return her beloved Muhammad Mustafa.

Turning, she discovers the babe, the Welfare of Humankind, deep in worship, facing the direction of the holy Kaaba in the position of full prostration. He now sits back and intones the prayer of witness to Divine Unity, *ashhadu anna la ilaha illallah,* pointing with the first finger of his small right hand. The babe cries aloud to Allah in heart-melting tones, "O Allah, I turn to You and to You alone. Bring me my spiritual community. Let all human beings achieve nearness to You."

The sweet Amina perceives this vision with such vivid clarity that she falls to the earth in prostration.

"My mind now returns to its accustomed plane of awareness, and the vision of the babe in prayer dissolves, revealing my precious son before me. But to confirm the miracle, his cord is already severed, his eyes are lined in dark charcoal, and his circumcision is already perfectly performed, although I am entirely alone in the room.

"His noble small face shines with a conscious smile, like the sunrise, that awakens my deepest love. My heart is overwhelmed with the fire of Divine Love. I clasp the holy child to my breast and weep. His lips are struggling to form words, not moving fluently as in my vision. With my head bent close, I can hear with wonder the delicate sounds, not baby talk but pure Arabic, as he submits his will to the Will of Allah and murmurs over and over again, *ummati, ummati,* my spiritual community, my people, all humanity."

It is for us, dearest brothers and sisters, that this newborn infant intercedes, praying that we will be well guided along the straight path of Wisdom, Truth, and Love. *My spiritual community!* He addresses us so tenderly. Let us offer him our loving salams in return, for to love the Messenger of Allah is perfect healing and perfect peace.

During this mystic night of the Prophet's birth, certain noble and upright citizens of Mecca, circumambulating the holy Kaaba while deep in worship and concentration, perceive through spiritual vision the entire House of Allah bow down to the descending soul of light and right itself again with every stone in place. These true worshipers turn to one another and proclaim spontaneously, "The Welfare of Humankind has finally taken birth among us." Their spiritual ears hear the holy Kaaba speak in triumphant tones: "Tonight is born a soul as radiant as the sun, who will bring healing and clarifying light to all humanity. He will cleanse My sacred space from all idolatry and lower forms of worship, purifying the delusion that imagines any sources of Life, Power, and Wisdom other than the One Source. His spiritual community, with purity of heart, humbly living the holy way of life, will circumambulate Me, barefoot, as Moses faced the burning bush, without covering their heads with any earthly veil."

Upon this mystic night of nights, the drums of the Jihad of Truth resound, and Satan tumbles from his false heaven of arrogance like a falling star. In the sacred and ancient House of Allah, elemental forces inhabiting the idols there are immediately driven out, unable to exist any longer in the atmosphere of purity generated by the descent of the Prophet's soul.

The Holy Person of the Prophet

The Glory of the World, when he reached the mature age of forty years, assumed the mysterious and heavy Crown of Prophecy, often hearing the Voice of Allah resounding in the depths of his secret heart: "O faithful servant, *al-amīn,* I have created you to be My absolute Mercy to the universe."

Priceless jewels of Quranic verses began to be revealed within the pristine awareness of the Prophet, accompanied by miraculous events beyond numbering. Never did that sacred body, instrument of Divine Revelation that transmitted the infinite treasure of the Holy Quran, cast any shadow upon the earth, even in the brightest desert sun. It was a human body so completely permeated with Divine Light that it became merged in Light, and light can cast no shadow.

Above the striking beauty of that noble head, a bright cloud was always seen in the sky, shading him with springtime coolness even beneath the burning summer sun and creating the refreshment of Paradise for those around him even when surrounded by the world of conflict and selfish desire.

The Prophet's holy eyes saw clearly through the surface of this world into both subtle and heavenly realms. His discernment of fragrances was so refined that whenever the sublime Gabriel began his descent with a verse of Quranic revelation, the distinctively sweet scent of the Archangel immediately filled the awareness of the Messenger, who would wait in ecstatic concentration for the noble message to arrive.

The perfectly shaped and sensitive lips of that most holy man needed only to move faintly and the sun and stars would tremble. When the cool breeze at dawn prayer stirred his shoulder-length hair, scents of musk and amber would fill the congregation. His teeth, like pearls, sparkled so brightly on moonlit nights that he became a second moon on earth. From his chest, beams of radiance shone forth, so his companions could walk safely even in the darkest night. The beloved one of Allah once divided the moon with a powerful spiritual gesture, using the right forefinger that witnesses to Divine Unity, as Moses divided the Red Sea with his staff. Thus, division can become a symbol for Divine Indivisibility. The Ruler of the World would often plant palm trees and pluck fresh dates within the hour, as the Mother of Jesus received the spontaneous fruit of her son from Allah Most High.

None can recount all the miraculous actions or describe the overwhelming beauty of the holy person of the Prophet, Pearl of the Universe, even were illumined singers of mystic songs to sing until the End of Time.

The Ascension of the Prophet

Come close now, true dervishes of love, to the very fire of Divine Love, which burns in the Prophet's noble heart as he ascends through heavenly spheres to the most intimate Presence of Allah. This same fire of love will now burn within you, for such is the highest mystic teaching of Islam.

Upon this Thursday night in Ramadan, this glorious Night of Power, our holy Prophet, the royal Shaykh dressed in clothes of a common man, rests in his humble dwelling in the cool of evening, immersed in the station of mystic union natural to his being. Allah now orders His sublime Archangel to prepare for the Prophet's ascension: "Take from Paradise, O servant Gabriel, the ceremonial belt of power, the jeweled crown of wisdom, and the chestnut steed of purity. Bring them to My beloved one on earth, that he may traverse the heavens and behold me without intermediary."

Gabriel enters the dimension of Paradise where countless beautiful horses of spiritual qualities are grazing, beholding in their midst one most noble steed with tears streaming from its tender eyes like a river in floodtime. This magnificent mount neither crops the rich green grass nor drinks from the pure streams of Paradise, for its heart is melted by the fire of love. The Archangel inquires, "Why are you weeping, noble one? What has wounded you so profoundly? What transcendent longing has separated you so severely from the natural pleasures of your fellow creatures?"

The dervish steed responds, "For seventy thousand timeless years of preparation, my only food and drink has been Divine Love. Since that moment willed by Allah when I first heard angelic beings singing sweetly and in harmony the holy name Muhammad, my state has been elevated from astonishment to spiritual bewilderment, till I can no longer distinguish left from right or day from night. The inconceivable bearer of that profound name, whoever he may be, is my Shaykh for eternity. My very heart has become simply the longing to gaze at last upon his face. This dimension of Paradise has become for me no more than a prison of sighing and groaning with the pangs of separation. Until I am united with the one whose glorious name I cherish, I will remain imprisoned in this unbearable longing."

With amazement and joy the Archangel cries, "O matchless celestial steed, languish no more. Allah Most High has elevated you to the station of companionship with His holy Prophet Muhammad, may supernal peace and the blessing of union always be upon him. Whoever bears in his body the fiery traces of mystic love will certainly know some day the inexpressible delight of union. Come meet your beloved friend, as Allah has ordained. Receive the healing balm of his luminous gaze and gentle touch."

By the Permission of Allah, this spiritual steed and Archangel Gabriel come instantly before the Culmination of Prophecy, the beloved Muhammad Mustafa. The Archangel bows low: "Most profound greetings, O holy soul. We bring you the loving salams of Allah, Who sends His most tender blessings upon you, upon your beautiful family, and upon your noble spiritual community.

"He invites you to ascend through the seven heavens, to become the lone guest of His transcendent Presence, and to gaze directly into His secret Essence. Tonight, O most blessed one of creation, Allah wishes to unveil the mystery, the Hidden Treasure concealed from all eternity, the very Face of Essence. Tonight, both Paradise and the eighteen thousand universes are flooded with Divine Grace as sweet and clear as the sacred waters of the spring of Zam Zam. Countless angels are ascending and descending, bringing the joyous news of your coming. The impassable gates of the seven heavens and the Paradise above them have tonight been thrown open wide. The richest blessings these realms contain are being scattered freely throughout creation. The abodes of spiritual bliss, in ascending order of subtlety and sanctity, await your passing. Please ascend with us, O noble Muhammad Mustafa, into the eternal heights. I bring you this steed of light from Paradise as a confirmation of the Divine Invitation."

At once the holy Prophet arises in complete submission to embrace the Will of Allah, girding himself with the belt of power, donning the crown of wisdom, and mounting the steed of purity—these gifts from Paradise perfectly natural and comfortable to him who is Paradise. The ecstatic mount spreads its chestnut wings and soars in spiritual flight, following Gabriel, who bears keys to all the subtle and heavenly dimensions. At his very next breath, the Sultan of the Nations dismounts in the ancient sanctuary of holiness, Jerusalem the Sacred. Here the luminous souls of the 124,000 Prophets come to kiss his hand. Guided by the direct Words of Allah, they form lines of prayer behind the Imam of all Imams and perform two cycles of prayerful prostration on the very site of the temple of Solomon.

Now the heavenly ladder of light appears. Leaping from rung to rung, from plane to plane, the beautifully mounted Prophet rides to the Presence of his Lord, observing the wonders of subtle realms along his way.

Suddenly, the mystical travelers glimpse the shining structure of the seven heavens beyond space and time. The holy Prophet, bright Bird of Paradise, enters with dignity and grace through vast gates, thrown open in ecstatic welcome. There he observes countless ranks of angels engaging in various modes of perpetual worship and meditation. Some repeat the beautiful Names of Allah; others sing mystic hymns to His Glory. Some accompany their silent praises with prayer beads like drops of molten gold; others perform the outward *dhikr,* whirling and calling out in ecstasy the powerful Names of the Most Sublime. Some remain immersed in ceaseless prostration, others stand perfectly upright, gazing at

Truth, and still others bow low in humble adoration. All are enraptured by Divine Love, lost in various degrees of mystic union.

These angelic beings turn with ecstasy from their contemplation and rush to welcome the Messenger of Allah, kissing his noble hand in the dervish greeting of love and congratulating him on his unprecedented spiritual journey. In perfect unison, with a single voice, they proclaim, "*Marhaba ya rasulallah*. Welcome to our eternal abode, O exalted Muhammad, our compassionate intercessor on the Day of Judgment. We perceive that the green turban of supreme realization has been conferred upon you this mysterious night. Please explore to your heart's full pleasure the ninety-nine levels of Paradise, and join us at the Divine Throne to hear sublime teachings from the Voice of Truth. You have received a spiritual honor never before accorded to a human being. Your worthiness and purity are unique."

On this luminous night outside time, the Prophet roams through gardens resplendent beyond description, led on and on by passionate love for Allah and embracing with his powerful gaze higher and higher spheres of meaning. From each sphere, he assimilates a treasure of esoteric wisdom, until his spiritual steps reach the very boundary of the Absolute. Beyond this frontier, even Archangel Gabriel has never ventured, nor can he venture there, remaining limited by his angelic nature.

Only humanity, to whom the angels bowed in the beginning by Divine Decree, can enter the formless Garden of Essence and behold the Essential Face. When Gabriel signals that he can go no further, the Mercy of Allah to all the Worlds declares in tones faintly touched by sadness, "Dear holy companion, we have always traveled together on Allah's mystic way. How can I go on without you into the realmlessness that is unknown and unknowable?"

At this the Archangel cries, "O Prophet uniquely beloved of Allah, you are a stranger nowhere. Divine Essence is your very homeland. For you alone has Allah All-Powerful created the universes and the heavenly spheres we have just traversed together, filled with diverse beings on more and more subtle planes of being. The human soul is the crown of creation, and you, O holy Prophet, are the Crown of Souls. You are the inheritor of the Essence of Allah. By the Will of Allah, I cannot conceive what lies beyond this most subtle boundary. You alone can venture there, where destruction would await me."

The Glory of the Universe replies, "Then stay within the borders of creation, as Allah has ordained. But I, whose steps are guided by Divine Love alone, must go forward, even if it means the dissolution of my created being. The human soul is commanded to abandon all and plunge into the Beloved. The only way to attain the goal of mystic Love is to lose one's head beneath Love's Own Most Merciful Sword."

While he and Gabriel are conversing, the essential energy of Allah takes form and invites the holy Prophet to follow. The Sultan of Lovers then disappears across the uncrossable boundary of creation. He now perceives only vast and formless depths—neither heaven nor earth nor figure nor structure. This boundlessness is neither full nor empty. Its nature is ineffable, incomprehensible, yet sublime.

Even within this boundless mystery, seventy thousand veils of light must be removed before the Original Abode of Essence is revealed. As each veil dissolves in turn, the majestic Voice of Allah calls from within the secret heart of the Prophet, "Approach, My beloved friend. Do not delay. I am longing to receive you."

Following this spaceless, timeless, formless journey, the Quality of Truth stands before the Essence of Truth. Essence blossoms into a six-dimensional rainbow of Divine Beauty. The holy Prophet perceives Majestic Oneness take formless form, clear as a full moon in the night sky, thus perceiving Allah as lovers of Reality will perceive Him in the highest dimension of Paradise.

Essence communes with the beloved Prophet in modes unimaginable, where neither word nor voice nor sign nor touch is needed. With Divine Tenderness far beyond even the Prophet's enlightened expectation, the Voice of Truth resounds through his entire being: "I am your Secret Heart, your True Desire. I am your Perfect Refuge, your Only Love. I am the Divine Reality you worship. For Me alone, O Muhammad of Light, you have sighed unceasingly throughout the long journey of prophecy—through Adam, Noah, Abraham, Isaac, Ishmael, Jacob, Joseph, Moses,

David, Solomon, and Jesus, and now as the earthly Muhammad of Arabia. You have secretly called out, 'Why can I not directly behold the beauty of the Essence of Beauty?' Come now, friend of all souls, enter My Essence. The love I feel for you is beyond My love for the whole creation. All humanity is your humble servant. Whatever you desire from the infinite treasure-house of Divine Power and Blessing is yours for the asking—thousandfold healing, thousandfold illumination, thousandfold love."

With balance and dignity, the heart of the Prophet replies to the Heart of Reality, "*Ya rahman, ya rahim.* O compassionate and merciful Lord, Who veils the faults of those who love You. O Forgiveness Absolute, bless the human beings who are wandering in search of love and truth. Show them a straight path into Your sublime Presence. So many are lost in destructive rebellion. I fear they will misinterpret Your supernal Light and thereby experience Your overwhelming Radiance as the fire of hell. O Most Majestic Allah, this is my sole petition: my spiritual community, my people, all humanity—may they be accepted by You."

From the Living Truth a cry of love resounds: "I grant them salvation in your name, sublime lover of humankind, O Muhammad, friend of Allah and light of guidance for all souls. My highest Paradise I promise for the spiritual station of those who love you. O beloved one, this is but the smallest of gifts. I offer you as well the essence of My Love, O gentle lover. Your mercy and compassion have turned to those in ignorance and rebellion when you could have asked for and received the whole of creation, including Paradise. You are indeed the mirror of My Divine Nature, *rahman* and *rahim,* Tender Mercy and Compassionate Love. This is why your name, O First Light, is inscribed with Mine even from before eternity as the mystic affirmation *la ilaha illallah muhammad rasulallah.*" Lover and Beloved now merge in a union not even mystical language can describe.

Allah Most Sublime continues to speak within the Prophet's secret heart: "Beloved Messenger, your soul will never turn from the experience of this essential union. Return now into temporality and invite humanity, the created expression of My Essence, to come unto Me, to gaze at last upon the Source of Truth and

Beauty. Traveling beyond time and eternity, you have interceded for them all at the Abode of Essence. I grant to your own spiritual community the Daily Prayers of Islam, which when performed with purity of heart will lift them into Paradise while still on earth. Every level of heavenly experience is contained in the Prayer. The Prayer will lead them along the very way of ascension that I led you. Those whom I empower to perform the Prayer five times daily will be in the spiritual state of those who pray fifty times daily—the state of constant remembrance of Allah."

During this conversation between Supreme Lord and supreme servant, though it lasts but a timeless second, all the vast inner teachings of Islam are transmitted in ninety thousand concise words. In that same breath, the Prince of Prophets and Sultan of Lovers is once more aware of his humble dwelling in the desert of Arabia. His noble bed is still warm at his return.

To awestruck companions, the Messenger of Allah now narrates this mystic night journey through eternity, and their minds, though pure and disciplined, reel with spiritual inebriation. These sublime companions of the Prophet, among the greatest adepts humanity has ever known, speak with a single voice to their consummate mystic guide: "O Light of Prophecy to whom we turn the face of our soul, blessed beyond all measure is your most exalted heavenly journey. Please accept us as the slaves of your love, O glorious Master of Love. In our secret hearts, you are the full moon of Truth, eternally rising. To be your companions, to be your spiritual nation, to be pleasing unto you, O Messenger of Allah, is our life and very being."

Salutations to the Holy Prophet

O Culmination of Prophecy, Muhammad Most Praiseworthy, Most Perfect of Allah's Creations, Caretaker of Divine Revelation, *al-amin,* Absolutely Trustworthy One, Purest of Arabs, Six-dimensional Mirror, Full Moon of Truth Shining in the Night of Divine Mystery, Bringer of Sublime Good Tidings, Sweet-voiced Nightingale of Love.

O Sultan of Hearts in the Garb of a Humble Man, Matchless, Priceless, Incomparable Pearl, *efendimiz,* Most Precious Spiritual Master.

O Glory of All Realms, Glory of Prophets and Messengers, *habibullah,* Most Profoundly Beloved of Allah, Perfect Guide along the Mystic Way.

O Living Presence of the Divine Mission, *hazrati paygambar,* Most Revered Prophet, One Who Ceaselessly Praises Allah, Gatherer of Humanity, Phoenix of the Egg of Pure Religion, *al-imam,* Leader of the Prayers, Imam of All Muslims, One Whose Beauty Kindles the Divine Words "But for you I would not have created the universe."

O Mystic Knower of the Sacred Law, Most Exalted One in Paradise, Seal of the Prophets Who Announces the Distribution of Divine Light to All Humanity, Seal of Revelation Who Announces the Completion of the Drama of the 124,000 Prophets.

O Best of Humanity, Shaykh of the Entire World, Master of the Temporal and Eternal Realms, One Whose Light Annihilates the Darkness of the Limited Self, Brilliant Gnostic Moon of Yemen, *mahbub,* Supreme Beloved, *mahmud,* Powerful Magnet of Praise.

O Kingly Sun of the Eternal Day of Knowledge, First Light Shining from the Divine Command *Be!,* World-illuminating Lamp of Truth, Life-giving Sun of *sharia.*

O Sage Always Wrapped in the Cloak of Deep Meditation, Muhammad Deserving Most Sublime Praise, One Who Equips Hearts and Souls for the Journey, Specially Selected by Allah, Truthful Speaker, Decider and Decision, *al-murshid al-akbar,* Incomparable Spiritual Teacher and Guide, *mustafa,* Divinely Elected One, One Who Is Wrapped in Divine Love and Protection.

O Prophet of the Holy Sanctuary at the Center of the Universe, Prophet of the End of Time, Prophet of Pure Divine Mercy, Prophet of Repentance Whose Intercession Is Always Successful, One Who Compassionately Warns Humanity, Center of the Nine Circles, Central Point of the Most Resplendent Circle.

O First Light, Light that Enlightens the World, Light of Guidance, Clear Light, Primordial Light, Culminating Light, Light of the Throne, Light of Lights, Light of Allah within the Light of Allah, Messenger of Messengers, Seal of Messengers, Complete Mercy of Allah to All Humanity, Messenger of Holy Warfare against Negation, Messenger of Sweet Solace and Comfort, Messenger to Human Beings and Beings on Subtle Planes.

O Resting Place and Sanctuary of the Prophetic Mission, Leader of the Advanced Guard, Holder of the Divine Explanation, Holder of the Green Banner of Infinite Praise, Upholder of the True Law.

O Fountain of the Energy of Life, One Who Is Crowned with the Mystic Green Turban, Prince of Prophets, Prince of the Universe, Master of the Righteous, Master of Endless Horizons, Master of Humankind, Master of Prophets, Master of the Two Worlds, Master of All Divine Emissaries, Master of the Mystic Masters.

O Living Lamp of the Way of Perfect Guidance, Sultan of Dervishes, Intercessor for Hopeless Sinners, Supreme Intercessor on the Day of Judgment, Intercessor for Every Spiritual Nation, Intercessor for All Humankind, King of Both Worlds, Perfect Witness of Truth, Caretaker of the Vast Expanse of Quranic Meaning, Caretaker of the Ocean of Love, Overseer of the Awesome Tumult of Resurrection.

O Watchman of the Night and Morning of Creation, Delicate Early Morning Sun of Wisdom, Source of Light for Men and Jinn, True Sharia, Perfect Model of the Holy

Way of Life, Prophet Beyond Letters, Unlettered Man of Truth, Living Quranic Sura Ta Ha, Living Quranic Sura Ya Sin.

O Quintessence of the Universe, Crest Jewel on the Forehead of Humanity, Rose in the Meadow of Prophecy, Diamond in the Crown of Prophecy, Central Pole in the Tent of Relative Existence, may the Peace of Allah and the mystic union with Divine Love exalt you eternally and exalt your noble family and your spiritual companions throughout time and eternity, gathering all souls together into the Culmination of Light and Love.

To be drawn into the Light of Allah and not be blinded, O brothers and sisters, repeat with longing and with love:

as-salatu wa-s-salamu alayka ya rasulallah
as-salatu wa-s-salamu alayka ya habiballah
Heartfelt and profound greetings to your sublime soul,
O incomparable Prophet and uniquely beloved one of Allah.

HEART OF THE HOLY QURAN This essay uses the mode of gnostic unveiling, which the great Islamic Shaykh of thirteenth-century Spain, Ibn Arabi, contrasted with the procedure of scholarly investigation and commentary. This approach of unveiling serves better than a more linear, rational commentative style to honor the noble Quranic chapter, Ya Sin, which many Dervish Orders consider the mysterious center and pivot of the Arabic Quran. The various sections of this composition came forth while fasting on successive days of Ramadan.

4
Heart of the Holy Quran
Unveiling Sura Ya Sin

Verse 1

Two mystic letters from the boundless language of Allah reflect in the clear mirror of the Arabic Quran as *ya* and *sin*. By these radiant and resonant letters is the beloved Muhammad, upon him be peace, known among the countless palaces of the angelic realms, which surround the visible cosmos as a vast, shining sea would surround a mustard seed. This Muhammad of Light known by the angels is the First Light, the transparent Pillar of Light shining timelessly from the core of Reality. This Muhammadan Light is the very longing of the inconceivable Essence, the Hidden Treasure to be revealed, to be known, to be loved. This metaphysical Column of Radiance is the supreme devotion of living light to the single Source of Light and Life. Before the eternal angelic realms are created as flawless, gemlike worlds of praise, crystalizing deep within Divine Light, Muhammad the Principle and Pillar of Light stands alone before his Lord.

From the moment they are called into being, the angelic emissaries of Divine Light intuitively know through the letters *ya sin* that prismatic First Light from whom, through whom, and for whom they were created by Allah Most High. "Ya Sin, Ya Sin, Ya Sin" they call continuously to the Light of Prophecy with the beautiful resonance of their dynamic play of light, created by Allah to be pure praise. Listen to these heavenly voices as they cry: "May the supernal Peace and Blessing of the Supreme Source pour forth abundantly upon *Ya Sin,* our beloved light, generating him as the five-pointed golden star, the brilliant crown of creation, which is perfect humanity, the mysterious meeting place of Divine Essence and Divine Attributes in the sphere of organic creation. *Ya Sin* is the Pen of Universal Intellect by whom the Quran of the eighteen thousand universes is written on clear pages of light that chant themselves as temporality and eternity."

Angels are created out of the primordial Muhammadan Light, their magnificent praise springing forth from his own original and incomparable words of praise, *la ilaha illallah,* there is nothing apart from Supreme Reality. So too are human souls created from this beloved Light. Affirmation of unity is our essential nature—the very Light of Knowledge and Light of Mercy, the very Light of Lights known as *Ya Sin.* We are the gracefully spreading rays of that First Light, that gnostic lightning flash, *la ilaha illallah.* We are microscopic rays of that vast Light of Guidance, *Ya Sin,* who has manifested through brilliant prophetic souls during human history, awakening billions of minds and hearts to the Light and as the Light. So with the angels we can equally cry to him, our beloved Light, "Ya Sin, Ya Sin, Ya Sin."

We call out not from the eternal angelic realms, which remain astonished by the mystical ascension through them of the Muhammad of Arabia, may peace be upon him. We call out from the blessed days of the fullness and completion of revelation. This ocean planet has now become one holy place of prayer, one mosque, its very dust sanctified by the Muhammadan sandals that stood before the Throne of Mercy and touched the unapproachable Ground of Essence. During this era of earth's history, we have been supremely privileged to witness the preeternal Muhammad, the Principle of Prophecy, become fully expressed at last as the Seal of Prophecy, the fragrant, beautiful, prayerful Muhammad of Arabia. We are complete.

Therefore, we can sing with the rich tones of the Holy Quran, in which the wisdom and beauty of all Prophets is conserved and fused: "May the Praise of Allah Most High and of His exalted angels be upon Muhammad, the perfection of humanity, worthy of the praise with which the angels first praised the transcendental Adam. And may Divine Praise and Divine Good Pleasure be directed as well to his beautiful family and to all his noble companions who will arise throughout history until the End of Time. These lovers will reflect the light of his perfect humanity in the bright mirrors of their hearts, cleansed and sanctified by the way of universal Islam, the way that fully releases daily life into all-embracing Divine Life."

With the ecstatic cry "Ya Sin, Ya Sin, Ya Sin," we have reached the glorious heart of the vast body of the Arabic Quran. The Mother of Books lifts her veil as a bride to her husband's gaze. Now we can contemplate perfect humanity as the Face of Supreme Love. The Light of Quran is the Light of Guidance, the original Muhammad of Light, who later came to be known as *Ya Sin* to angels and finally to human lovers living in the completion of revelation.

This Muhammadan Light is the pulsating Heart of Quran. The Muhammad of Arabia, the Mercy of Allah to all the Worlds, not only came bearing the Glorious Quran but is the Glory of the Quran. He both received Quran and demonstrated Quran with his very being. His movements were Quran. His life in all detail was Quran and is Quran. His human face shone like a bright page of the Holy Quran, illuminated with pure gold. Thus the foreheads of his noble spiritual family, the true lovers, shine with precisely this same Quranic radiance and the mystic letters *ya* and *sin*.

Lead us, O *Ya Sin,* through the central door of Sura Ya Sin, to encounter the mysterious illumination of *Ya Sin* everywhere, through everyone, in every breath and every event of creation, which all reflect the beautiful Divine Names of Allah Most High. *Ya Sin* is the secret. Sura Ya Sin is the key to the secret.

Glory to the Lord of the Worlds!
amin amin alhamdulillahi rabbi-l-alamin

Verses 2–12

Creation exists for the sake of the beloved Muhammad of Light, who shines through and as the Glorious Quran, the Mother Book that bears all the Wisdom of Allah within her radiant womb. Since the beginning of human history, this Divine Wisdom has flowed as a single stream of revelatory energy through 124,000 noble Prophets, including 313 Chosen Messengers, may peace be upon them all, never repeating mechanically but remaining ever unique, alive with Divine Life, free from limited human conceptions.

Allah Most High confirms that the Muhammad of Light, the original light and mercy of the countless realms of Allah's Light and Mercy, the exalted *Ya Sin,* stands within this inner circle of 313 Messengers of Love. This is the central circle among billions of concentric circles of precious human souls. More elevated than angels, these souls face only the Truth, magnifying their Lord and merging with their Lord. We witness with our whole being that *Ya Sin,* the First Light, is the Shaykh of these 313 incomparable beings of radiance, each of whom has established, by the Will of Allah and by the most heroic personal effort, a true and complete spiritual way for their own nation, a way that leads through the veils and metaphors of Light into the Source of Light.

At the center of the sublime circle of 313 stand the 12 magnificent spiritual leaders of the Prophetic Body, upon them be Divine Peace: Adam, Noah, Abraham, Ishmael, Isaac, Jacob, Joseph, Moses, David, Solomon, Jesus, and the Muhammad of Arabia. Theirs is the Straight Path, the Direct Way, the Only Way. The single Divine Message that they bear manifests at the heart of temporality as the universal Call to Prayer, the universal way of living prayer. This Message remains always one in essence yet is uniquely expressed through each prophetic luminary. This is the universal Revelation, descending from the radiant black rain cloud of Divine Essence as a downpour of fragrance and refreshment for all creations that are called forth by one complete, perfect Creativity. This rain of revelation drenches the universe from end to end.

Exalted beyond any conception or imagination, beyond any measure or symbol, is Allah All-Powerful, and yet His Mercy is more extensive and mysterious than His Power. Every drop of this temporal and eternal rain of His Holy Mystery is a Book of Revelation, a Book of Reality. Remembering Allah ecstatically with every breath, we stand beneath this wonderful rain of living gems, this cascade of revelatory verses and clear signs.

To prepare and open the soil of the heart for the rain of Divine Grace, humanity must be warned and chastened, disciplined and refined, awakened from the confused dream of carelessness, made sensitive and responsive to every moment and to every breath of life as the very flow of Divine Life. Carefulness, caring, concern, and intuitive recognition of the mystic waves of the Divine Will, as one recognizes instantly the face of a beloved friend—this is the prophetic call, the prophetic process. This is the highest training of humanity for which the noble Messengers come and for which their families, companions, and sublime lovers sacrifice their lives from age to age. The Word of Allah, manifest through all inspired scriptures, always bears Truth, demonstrates Truth, and awakens the innate human ability to recognize Truth, unerringly and uncompromisingly.

Divine Words, streaming from the Prophets and from the mystic guides who inherit the spiritual wealth of Prophecy, alone can humble the proud. Those whose entire mode of being has become rigid and awkward through pride are isolated by Allah Most Wise from the delicate and delightful realms of holy vision, lest they disdain or even despise spiritual beauty, bringing even more harm upon themselves and upon other precious human souls who may associate with them. These arrogant ones are mercifully surrounded by Divine Barriers and Divine Veils, lest they gaze upon the slightest spiritual secret with an impure heart. Their habitual, self-perpetuating negativity is impervious to the subtle nature of the prophetic radiance, the Light of Guidance, which is streaming into the universe at this very moment through the vast channel of *Ya Sin*.

Only those whose hearts are melted by the sweet fire of love, whose eyes are purified by tears of love, and whose minds have become utterly humbled through constant contemplation of the awesomeness of Allah—only lovers of the Great Love, lost in love and found in love, can receive the message of *Ya Sin* directly from the most tender Source of Love. These are the ones who experience total purification, total Divine Forgiveness for the slightest shadow of negation or rebellion, hatred or cruelty, that has ever touched their minds, even for an instant. What inconceivable joy is kindled by the news of this Resurrection of Love in a body and mind composed entirely of Divine Love, not waiting for the End of Time but standing up here and now from the narrow grave of the limited self. This bliss beyond measure alone can experience and express the Generosity of Allah.

Whoever longs for supreme bliss may now come forward to receive. This is the astonishing invitation of Allah, transmitted today through His sublime lovers and intimate friends, who are given the Divine Power to awaken and resurrect dead hearts—both those who have died spiritually, while continuing to live on this earth in the bondage of self-centeredness, and those who have died physically and may have entered the terrible grave of despair and tortured conscience.

The Call to Divine Life, which streams as a vast torrent through the very being of these universal lovers, can bring the living and the dead to higher and higher spiritual levels. This evolutionary change of levels is demonstrated unequivocally by authentic dreams and visions. The transformation and elevation of human beings who have strayed from the Truth, accomplished by the healing touch of light-bearing lovers of Truth, is promised and confirmed in the Holy Quran, shining forth with perfect clarity from its central chamber, Sura Ya Sin. Calls out the Divine Voice through this Sura, "You will have the power to quicken and resurrect every heart who has become divinely attracted to you and who has been joined subtly to you by Allah All-Merciful—companions, known and unknown, from the present and from the past." This is the secret promise to all the lovers of *Ya Sin,* the First Light from whom the souls were formed.

For this Divine Promise of becoming the Mercy of Allah to human hearts, the noble Messengers and their successors weep with profound gratitude. Every heart is an entire life-bearing world, an entire planetary civilization, precious beyond any calculation. *Ya Sin, Ya Sin, Ya Sin!* Attract all hearts without exception to the prophetic light. Heal, elevate, and illumine them. Fill them with universal love. Grant them the purest peace of Oneness, the highest station of Islam.

Glory to the Lord of the Worlds!
amin amin alhamdulillahi rabbi-l-alamin

Verses 13–32

The universe is a book of parables. Human history is an interwoven teaching story. Every prayer time marks an advancement in spiritual knowledge for persons who see with the eyes of their hearts wide open. Companionship among lovers of Truth provides the most radiant explanation, which contains subtle levels of understanding that unfold as this companionship deepens, as intimate association becomes union, just as grapes growing together on one potent vine are crushed and fermented into the Wine of Love.

The most exalted experience of spiritual companionship arises among the apostles of a world teacher. These blessed souls, wherever they appear in history, form a single, transtemporal mystical body of the Prophetic Message. They serve humbly and lovingly as limbs and organs for one another, trembling like green leaves in a bright wind as the breath of the Holy Spirit passes among them, causing them to dance and glisten in the sunlight of Truth.

Storms of resistance and persecution also sweep down upon the mystical body of prophetic companions. Yet the roots of the

Prophets, may peace be upon them all, are plunged deep in the soil of the secret heart, so the trees of their spiritual communities are never uprooted, even by the most bitter gales. The noble trunks, the great limbs, and even the smaller branches are fully alive with the sap of ecstasy, of conscious union with Reality. They bend gracefully under the blast, without breaking, while brittle worldly structures fracture instantly.

The Tree of Revelation whose root is the beloved Muhammad of Arabia, upon him be peace, has experienced fourteen centuries of vigorous growth. The Tree of Revelation planted by Allah Most High in the heart of Muhammad's intimate spiritual brother Jesus, upon him be peace, has experienced twenty centuries of vigorous growth. Some two billion precious souls living on the planet today are protected and refreshed beneath the life-giving shade of these two lofty and wide-spreading trees of sweet spiritual companionship, two trees from the same soil, whose limbs and roots touch and intertwine.

The intense lovers who form the mystical Orders of Islam have taken the beloved Jesus as their spiritual friend and model, emulating his mildness, his poverty, his simplicity, his dedication, his love, and his mystical union. With the eyes of the heart, the lovers of Islam clearly perceive the spiritual equality of these two greatest embodiments of Divine Love, Jesus and Muhammad, upon them both be perfect and abundant peace and upon their devoted companions throughout time. When the Prophet of Allah was cleansing the holy Kaaba in the City of Divine Power, clearing away primitive idols, he allowed the beautiful fresco of the Virgin Mary holding the child Jesus to remain, gracing the inner wall of the inmost sanctuary of Islam, just as the Virgin and her radiant child remain central to the Quran. The Holy Quran extols the ancient Christian fathers and mothers of the desert, who lived in the wilderness of total renunciation in order to live within Divine Life alone. The Holy Quran predicts that these renunciants will shed tears of love upon recognizing the authenticity of the bearer of the Quran.

There is a modern Christian lover of Truth whose eyes once overflowed with tears when he prayed the sublime Friday Prayers of Islam with dervish brothers and sisters. When he came forth from the Mosque of Delight into the most dynamic city of the modern world, he perceived more vividly than with physical vision a vast blue lake, right there among tall buildings and busy streets. Along the shore of the lake were strolling joyously, arm in arm, the two culminating prophets of Divine Love, demonstrating the perfect harmony of their spiritual ideals and calling for harmony among their spiritual communities. Their mystical bodies represent thirty-four centuries of cumulative experience, both gnostic wisdom and tender compassion.

Sura Ya Sin, heart of the Holy Quran, contains the story of the blessed companions of Jesus, upon him be peace, who enter the city of Antioch, which was the center of worldly dynamism in their cultural era. The wise and tender voice of the Light of Guidance mysteriously speaks through an unknown man of Antioch in support of these Christian apostles. Thus Jesus and Muhammad, may the most sublime blessings of Allah always be upon them, are inseparably joined by Sura Ya Sin at the heart of the Quran, at the heart of Islam, at the heart of the Drama of Revelation, beside the blue lake, the Fountain of Love in the Garden of Paradise, distributing freely to all humanity the ultimate drink of Divine Love.

Sura Ya Sin reveals that the awakened companions of Truth must enter the city of world power and transmit the message of love, which will cleanse that center from all traces of worldliness. The response of the city dwellers is always anger and terror, as an anxious child might feel toward the doctor who cleans and cauterizes his wound. "Truly, we have been sent on a mission to you," cry out the light-bearing apostles of Jesus to the anxious children of Antioch. "Truly, we have been sent on a mission to you," calls the resonant Arabic text of the Glorious Quran. With this directness and simplicity alone can Divine Love speak—only with this openness, with this vulnerability. When the Mission of Love is inevitably rejected by limited selves and limited societies, the love-bearing apostles, accused of deception, most humbly reply, "Our Lord alone knows that we have been sent on a mission to you, and our only responsibility is to proclaim His Clear Message."

Allah alone is the Knower. Allah alone is Clarity. Allah alone is the Power of the Mission. To this merciful Power, manifest through the shining eyes of the companions of Love, the tyrannical pharaoh of the grasping and domineering self cries out, "If you do not desist from teaching, we will stone you and inflict grievous punishment on you." The truly believing heart, the sultan heart that resplends with the nobility and majesty of faith, now shines forth in the divided city of human consciousness. This pure heart always exists, hidden within the city of the world, secretly manifest as the human soul, the crown of creation. This true heart, this perfect humanity, comes forth with deep dedication and runs toward the spiritual goal. The mysterious words of Sura Ya Sin continue to open before the gnostic eye: "There came running from the farthest part of the city a man crying, *O my people, obey the apostles.*" These are the sweet words of faith that continuously stream forth from the interior chamber of every human heart, even the heart of the Egyptian Pharaoh, although Allah seals and hardens the surface heart of negative beings to protect them from their own negativity.

This absolutely faithful witness, deep within the very structure of human consciousness, crying out from the radiant pages of Sura Ya Sin, presents this criterion for the authenticity of an apostle: "Obey those who ask no reward of you and who have themselves truly received guidance."

The authentic companions of Divine Love want nothing and need nothing from any being in creation, since they are nourished and protected by transcendent Reality alone. Thus they are supremely generous with their very lives to all conscious beings. They would gladly welcome martyrdom, becoming wheat ground up by the teeth of wild beasts and baked in an oven of furious flames, in order to become the bread of life to all those who are hungering for substantial spiritual teaching.

The bearers of Divine Love, whose very bodies and minds have become Divine Love, are those who have received the Light of Guidance into the core of their being. These representatives of the Prophets shine with Divine Light in every moment, not simply through their verbal teaching but in their own unshakable state of being. These fully surrendered servants of Divine Love are the most sweetly reasonable and yet the most uncompromising of all persons. Cries out the prophetic voice of *Ya Sin* in luminous words, "It would not be reasonable, nor would my being be harmonious, if I did not serve with my whole being Him Who created me and Him to Whom all of you will be brought home again. Shall I adopt idols or derivatives when I have Him, the Original Reality? I accept wholeheartedly whatever the Most Gracious One intends for me, whatever adversity, for it would be profound spiritual error to invoke derivative powers to protect me against the Will of the One True Power, Who is the Supreme Lord of every one of you listening to these words. My heart is overflowing with the royal treasures of faith. Freely receive, here and now, the wealth of the Prophets."

Then Sultan Allah, the King above earthly kings, calls to His perfect slave of Love, who is *Ya Sin,* the single Prophetic Light that shines through all the Messengers, "My beloved friend, enter the Garden of Essence." *Ya Sin* is completely overwhelmed by the mystery of Divine Mercy and weeps, "Would that all humanity could know what I know." For these sublime tears of universal compassion, for this most tender solidarity with human longing, *Ya Sin,* the Muhammad of Light, has received the Divine Promise of total forgiveness for every heart that is joined to him in love, by even a grain of love for him or for those who love him. This all-powerful attraction of Divine Love is his only army, his only weapon.

The cosmic explosion of limitless love, experienced whenever we turn our being entirely to the Lord in prayer or contemplation, reduces the resistance and recalcitrance of the world to ashes, like a madly blazing fire suddenly quenched with water. This sudden silence and stillness is the Day of Truth, when all consciousness stands naked and transparent before the Living Truth alone. The flames of mockery at the Truth, in which all limited selves indulge, are now mercifully dispelled.

O *Ya Sin,* be with us now and on that Final Day beyond time, which may dawn for us at any moment. Be our gracious intercessor, our clear witness, our inward completeness that needs nothing else. Be our beloved light of guidance.

Glory to the Lord of the Worlds!
amin amin alhamdulillahi rabbi-l-alamin

Verses 33–50

All creation is a sign from Allah Most Wise to His beloved souls, whom He called into being from the preeternal Muhammadan Light and whose very being is the primordial *Yes,* spoken by every soul without exception before entering the drama of temporality. This original acceptance of Allah's invitation to the Path of Return constitutes the essence of human consciousness. Therefore the true life breath of the soul sings ceaselessly, "Yes, I am coming, Gracious Lord. I am coming."

Creation is revealed as a reservoir of spiritual meaning to the soul who is aware of its own essence and is therefore consciously and constantly oriented toward the Source of Creation. The cycles of organic life on this earth are potent signs from the Source of Wisdom, signs that prefigure seasonal changes in the life of the soul, for each soul is like a life-bearing planet, revolving about the Sun of Truth along the invisible orbit of Allah's Will. The Holy Quran reveals that as the earth, which dies in the winter, is renewed in the spring, so the earthly body, which dies and falls into the soil as a potent seed crystal, is regenerated by Allah All-Powerful in the spiritual springtime and recrystallized as a miraculous body of pure radiance on higher planes of being. In this way, every expression of organic life is a sign and type for the future life of transcendence. This mysterious future life actually begins the moment we turn consciously toward Truth and enter the Path of Return.

The grain springing from cultivated fields prefigures the abundant harvest of Divine Teaching. The orchards of fruit-bearing trees, carefully pruned by the gardener, are seen with the eye of gnostic vision to herald the nourishing and refreshing fruits of Divine Wisdom, tasted by the illumined intellect after careful spiritual discipline. Vines heavy with grapes symbolize the abundance of spiritual emotions within the heart, which are crushed and fermented by Divine Power to become the Wine of Love. Fountains of clear water welling up from invisible springs deep within the earth symbolize the uprushing currents of spiritual energy in the subtle body, providing true refreshment for humanity, both in time and in the timeless realm called Paradise. Thus the whole creation is a window of revealed metaphor for souls to gaze through into the wonders of their spiritual homeland. To the Origin and Goal of the universe, souls are constantly returning; the entire cosmos is ecstatically returning. Not just a window to view the astonishing play of Divine Attributes, this transparent creation of Allah can become, at the depth of prayer or inward remembrance, an open door through which we can walk into Essence. Not even angels can follow.

Nothing in creation is essentially man-made, for our hands and minds are simply instruments of Allah. Nothing can be more awesome and illuminating than creation when perceived, in all its intricate detail, as Divine Teaching, Divine Manifestation, and Divine Call. Creation, when experienced as being in intrinsic communion with its Creator, is sheer gratitude. Manifest Being gives thanks with every breath, every heartbeat, every pulse of energy. The very function of the creation is thanksgiving.

Holy Quran reveals that the harmonious and mysterious

communion between male and female principles, each uniquely powerful yet fully complementary, is a mirror for Divine Wholeness, manifesting as wise and tender Divine Creativity. Holy Quran reveals that the simple rhythm of day and night expresses the dynamic of consciousness—the evolution of spiritual degrees, states, and stations within the soul. Holy Quran reveals that the precise course of the sun through the sky and the subtle phases of the moon bear secrets of knowledge, illumination, and mystical love for the contemplative who perceives creation with the eyes of inner wisdom. Creation is the intimate embrace of the Creator. This Divine Creation is neither material, mechanical, nor impersonal—not a prison or an outer darkness alienated from Divine Life. The Glorious Quran reveals that Allah alone is the Light of both heavenly and earthly realms. There is no darkness in Reality, Who is all Light, no deadness in Reality, Who is all Life, no impersonality in Reality, Who is all Mercy.

The teaching signs of Allah, scattered so profusely throughout creation, become most intensely focused in human history as the wonders that Divine Love performs through the beloved Prophets, may peace be upon them. Among the central prophetic signs is the bearing of the human race and all created life in the mystically spacious Ark of Noah—purified human consciousness become a vessel of salvation and regeneration for all beings from the terrible flood of arrogance, divisiveness, destructiveness, rebellion, heedlessness, cruelty, hatred, and indifference. Through the skillful and dedicated hands of the beloved Noah, upon him be peace, Allah Most Merciful created the Ark. Through human instrumentality, Allah is continuing to create this holy Ark from age to age—the vessel of true refuge, constructed by the inward certitude of faith, impervious to the storm and deluge of deluded and distorted human behavior. Were it not for the supremely merciful Will of Allah, humanity and the life of the entire cosmos would have succumbed again and again beneath waves of rebellion and negation. May human beings awaken to the dangers of this flood of divisiveness and come consciously for refuge to the Ark of Oneness!

Human beings constantly turn away from the Truth of Unity, not just through criminal actions or metaphysical denial but in the casual, careless, indifferent moments of their existence. O *Ya Sin,* O lightning flash of the First Light, awaken us moment by moment! May we become true dervishes, children of the Divine Moment, in which the Garden of Nearness flowers.

The waves of rebellion and insouciance can be extremely subtle. Persons who have denied the faithfulness inherent in human nature, beautifully designed by Allah Most High to become the crown of creation, avoid and ignore the smallest act of human kindness, such as feeding the hungry and regarding all humanity as one family. Desperately, these persons who have rejected the sanctity of their own personhood and the precious personhood of others turn away from and even try to escape the Signs of Allah, only to be confronted again and again, wherever they turn, with the demonstrations of Divine Love and by the lovers of Divine Love. They continue to keep their rebellion alive by asking absurd questions in the name of rationality and by demanding crude empirical explanations of the most refined spiritual teachings.

Allah All-Powerful will bring this tragic, empty argumentation to an end with a crescendo of the very Divine Resonance that already resounds harmoniously through all the revealed scriptures of humanity. This overwhelming Resonance and Living Word of Allah will seize human beings with the full force of Divine Love and will consume their self-distraction and self-importance. This is the experience of death, which comes to everyone, tearing some away most painfully from their chronic state of egocentricity. Yet mystic lovers long to die before they die, to stand up in the Resurrection of Love, here and now, on this green expanse of earth, which Allah All-Merciful has spread out for humanity like a rich prayer carpet, beautiful and balanced in design.

These are the Divine Words of Sura Ya Sin, full of countless levels of meaning and powers of healing, being spoken by the Voice of Allah within the secret heart of humanity.

Glory to the Lord of the Worlds!
amin amin alhamdulillahi rabbi-l-alamin

Verses 51–67

The magnificent crescendo of Divine Resonance will sound forth at the End of Time. This wave of awesome spiritual energy, like the clear call of a trumpet, will wash over the entire universe, bringing it to complete stillness and plunging all conscious beings, on all planes of being, deep into ecstatic trance. From tombs of limited awareness, beings of awareness will rush forth into limitless awareness. Persons who failed to practice prayer, worship, and meditation will at first be anxious and even anguished as they experience the flight of their fluid, formless bodies of radiance through the open space of Allah's Own Radiance. But their awareness will become centered and harmonious again when angelic and saintly guides, surrounding them as brilliant clouds of witnesses, remind them repeatedly, "This is the Day of Truth, which the All-Merciful One has promised in all the revealed scriptures of humankind."

Those resurrected ones who lived on earth in loving relationship with any authentic body of the Prophetic Message will be overjoyed to realize at last that the words spoken through divine inspiration by the apostles and lovers of the One God were accurate and true. Borne at speeds greater than the speed of thought on this single wave of Divine Resonance, every person from the beginning of history will be brought, in a timeless body composed solely of radiance, to the awesome Throne. These souls will encounter, with deep trembling of their whole being, the unimaginable and utterly astounding Divine Mercy, Love, Tenderness, and Compassion. Not a single soul will be wronged but cleansed, healed, and elevated according to every action or intention of goodness, kindness, generosity, reverence, and gratitude, no matter how minute, that occurred throughout its entire career on earth. By the light of these moments of affirmation, vastly magnified by the All-Merciful One, the shadows of negation will be dispelled.

Into this day outside time, all persons mysteriously awaken at the moment of physical death. The whole universe will awaken here at the moment of its physical dissolution at the End of Time. On this day, the companions of the Garden, those souls whose earthly lives have been watered with gushing springs of prayer and have borne the fruits of wisdom, will awaken with their entire being become sheer bliss. These companions of Divine Love will find themselves enthroned with royal dignity, experiencing the delicate springtime coolness of pure contemplation, surrounded by the ripe fruits of every mystical state and station—a profusion of spiritual colors, tastes, and fragrances directly revealing the wonders of their Lord. Whatever visionary experience or gnostic realization they long for, these souls will instantly receive. The Divine Word that consciously surrendered souls will hear emanating from the source of this Garden and pervading all the levels of subtle experience there will be "Peace, Peace, Peace! May My Own Peace be upon you," as Allah Most High sends each soul His loving greetings and salams, rich with the most abundant tenderness.

A glimpse of this Divine Tenderness can be experienced on earth through the purest mother's love. Therefore the Prophet Muhammad, upon him be peace, teaches in his noble Oral Tradition that the mother is the most important person for souls who long to live in the mystery of Divine Mercy called Islam. "The mother, the mother, the mother," he proclaimed in sweet tones, "and then the father." When the subtle body of our ideals and intentions stands before the motherly Throne of Mercy, all negativity will be clearly revealed as a terrible twist in human awareness, as the distrust and even the hatred of love. To the souls who have distorted themselves by the rejection of love, Divine Compassion clearly proclaims their spiritual error.

May we listen to the awesome Voice of Allah pulsating through the Quranic heart, Sura Ya Sin, addressing those souls who have tragically allowed the garden of their precious capacity to love to become choked with weeds or to wither from thirst. These Words of Judgment provide the shock of awakening necessary for those souls who were not awakened during their earthly lives by the beautiful Call of Allah through His beloved Prophets or even by the disillusioning experiences that Allah Most Wise allowed these careless ones to bring down upon themselves. These words of Divine Chastening impel such souls at last to enter the path of purification, now become excruciatingly painful because no temporal distractions remain to veil the searing sense of separation from the Source of Love.

The Supreme Beloved cries out, "O My souls who are consumed by your own negation of love, the Day of Perfect Love is dawning, and you must remain separate and apart, imprisoned in the negative intentions and actions by which you have deformed your spiritual body. Did the Source of Wisdom Who is now speaking not remind you repeatedly through My beloved Prophets, O children of Adam, that you should not in any way worship or cooperate with the subtle force of disintegration and disunion called Satan? The satanic principle of arrogance and rebellion has always clearly been the enemy of your balance and integrity.

"Who could be more adorable and worthy of worship than the Source of Love, the boundless Ocean of Love, always in waves of ecstasy, toward which all the various rivers of human love are racing, either unconsciously or consciously? The path of love along which souls of love return into the Source of Love is the Direct Way called Islam, the Essential Way that shines through the Prophets who have been sent to all nations.

"O children of Adam, so many of you have been led astray from My Path of Love by the principle of self-centeredness. Now your chronic fear of losing the limited self in Divine Love must inevitably experience My Radiance as the flames of hell. O rebellious ones, your primordial nature, which is pure acceptance, has become twisted into patterns of rejection. Allah Most Merciful is not rejecting you and can never reject you. You have rejected His Mercy by your very mode of being. None of the mere words of your mouth can intercede for you. No brilliant verbal evasions, no sophisticated philosophy, can save you, for the Source of Power Who is now speaking has sealed your lips. Instead, your very hands and feet—all the limbs, organs, and senses of your spiritual body—will witness clearly to the quality of your lifetime. Look carefully at yourselves! Where is the delicate beauty of love that is natural to your humanity, created to be a clear mirror of My Divine Love?"

Thus the terrible Word of Awakening will sear those souls who have indulged in negation but will not obliterate their faculty of spiritual vision. The Path of Return will still unfold graciously and invitingly. The innate spiritual longing of these souls will not be immobilized by Allah Most High. The original intention to return, spoken with sincerity by all souls before creation, will gradually dispel the veils of negation. Raging flames will melt into fragrant flowers of submission.

This is the mystical revelation of Sura Ya Sin, heart of the Quran where only love can gaze, where only love can enter, where only love reigns. Only love. Only love.

Glory to the Lord of the Worlds!
amin amin alhamdulillahi rabbi-l-alamin

Verses 68–83

The Resonance of Allah can infuse any language, miraculously reconstituting and resurrecting these historically evolved human words so they actually become the Divine Word. Therefore, the revealed scriptures of humanity bear the Light of Allah, their verses replete with secret dimensions that contain all Truth. Divine Radiance can similarly awaken, refine, and totally enlighten historical human beings, freeing them from limited perspectives, infinitely expanding their breast and enabling them consciously to embrace and express the Attributes of Allah. This transfiguration of human beings and human languages by Divine Light is the Drama of Revelation, the perfect communion of awareness between the Creator and His creation, a communion that could never be achieved, or even conceived, by any creature on its own.

The religions of humanity, with their authentically revealed Books of Light and their authentically inspired Prophets of Love, are essentially Divine Initiative, not independent human efforts to know Reality or please God. Allah All-Merciful is seeking the heart of humanity as a lover desperately seeks the beloved. The Creator's Longing, the Creator's Search, the Creator's Love is then reflected as a living mirror image in the heart of the creature and becomes human longing, human search, and human love. Allah alone is the source of the thirst and yearning of our whole being for the Source of Being, causing us to seek meaning and joy through every metaphor and veil of relative existence.

This ecstatic longing can rest only in its true homeland, the Garden of Essence. The circuit of love then becomes complete, as the soul of love returns to the Source of Love. Love pours into love, races into love, expands into love, and finds only love. Human love then realizes that it has always belonged completely to Divine Love. Divine Love created human lovers and then fell in love. Divine Love taught prayers and then prayed these prayers. Divine Love revealed scriptures and then contemplated these scriptures. Now the mirror image of human love disappears again into Original Reality. There is only One who perceives, only One who prays, only One who loves.

Poetry is metaphor, bringing separate dimensions together to illuminate each other. Divine Revelation is not metaphor. The Divine Knower knows no multiple dimensions. Holy Quran essentially strips away poetry from the human psyche, shining forth as universal Oneness, as naked Clarity without adornment or complexity. May we listen with awe to the Voice of Allah thundering within the dark blue rain cloud of Sura Ya Sin:

"The Ultimate Reality Who is now revealing this Arabic Quran has never guided My beloved Muhammad in the way of poetry, for metaphorical speech is not appropriate for the bearer of Divine Words. The Radiance and Resonance that manifest through him is nothing less than My Message of Clarity, My Radiant Quran, which illuminates the entire universe with the Light of Unity. This Glorious Quran constitutes warning and guidance for all persons until the End of Time. This Clear Quran presents an incontrovertible demonstration of the One Truth against all the limited selves in creation that reject or distort the Truth of Oneness. This Powerful Quran fully opens the eyes of the heart to the Source of Power Who is now speaking and Who alone provides every form of sustenance that human beings gather on this abundant and beautiful planet. My constant and overwhelming Divine Generosity cannot be perceived with ordinary vision. Only awakened vision, the gnostic eye opened by the touch of My Quran, can know true gratitude to Allah Most High. The clouded vision of humanity perceives My Creation as multiplicity and therefore acknowledges and even worships various sources of power in the universe, foolishly ignoring the Single Source of Power Who alone sustains and offers tender assistance to all lives. These disoriented persons who wander in multiplicity must eventually be confronted directly by the One Power. This experience will shock their habitual mode of perception.

"My beloved servant and Messenger Muhammad, do not become saddened by the low level of understanding manifest in humanity. The All-Merciful One Who is now speaking within your secret heart clearly knows their secret hearts of sincere longing and love. Yet how absurdly arrogant human beings become! Forgetting that their earthly forms were evolved by the Source of Power from mere drops of earthly substance, they now exalt their earthbound vision and their earthbound understanding and become adversaries of the transcendent Truth. Indulging in various superficial arguments against Allah's Revelation, they cry out, 'How can any power resurrect bones that have turned to dust and fill them with life again?' My beloved Muhammad, the Source of Wisdom will now reveal the very Divine Words that you should faithfully transmit to open the limited, conventional minds of those who question how the Source of Life can confer His Own Life upon all that lives. My beloved Messenger, proclaim this about the universal resurrection:

"'The One Power Who creates the entire universe on the temporal plane of being can certainly recreate this universe on the eternal plane of being. Boundless Creativity is the very nature of the Supreme and Only Reality. Just as you are not amazed when you see a green tree, cut down and kindled, transformed into a brilliant and dynamic fire, neither should you be amazed when you see the green tree of this earthly body transformed into the golden fire of the Body of Resurrection, composed of Divine Love alone. The One Power effortlessly creates Being as a symphonic flow of seven ascending planes of being. The entire physical plane, containing millions of inhabited planets, is no more than one seventh of Allah's Creation. Is this supremely creative Divine Power not capable of rekindling your earthly being on another plane of being? The Lord of the Worlds simply commands *Be!* and human beings spring into being on higher and higher levels and stations of being, until they cross over the far boundary of Being, venturing beyond Being into Essence, where even Archangel Gabriel cannot follow.'"

Glory to the One in Whose Hands of Power and Presence every event is always peacefully at rest. Glory to the One Who is tenderly drawing every life into His Own Life. These are the Words of Allah Most High, spoken into the secret heart of His First Light and Final Messenger, *Ya Sin,* as the radiant Arabic verses of Sura Ya Sin, heart of the Holy Quran and heart of Reality.

Glory to the Lord of the Worlds!
amin amin alhamdulillahi rabbi-l-alamin

ISLAMIC MEDITATIONS These twenty-six meditations were composed by the mode of unveiling, one each day, while fasting during Ramadan. They were read aloud to the community of the Masjid al-Farah after breaking the fast each sunset with a sumptuous meal. The gradually increasing intensity of these compositions traces the experiential contour of the ecstatic month, which celebrates the descent of the Arabic Quran. The simple words from the lips of the Prophet at the end of each meditation were the seeds from which each meditation emerged. These renderings of the prophetic utterances are on the level of *tafsir,* or interpretation. The apparently simple words of Quran and Hadith, the Oral Tradition of the Prophet, contain infinite levels of meaning, inexhaustible resplendence and resonance. Upon these revealed and inspired words alone, and not upon Islamic rational philosophy, Sufism remains firmly based. Even a small selection of these words, historically authenticated and supersaturated with the energy of Divine Blessing, are more powerful than entire treatises by scholars, ancient or modern. From these living words of Quran and Hadith alone emerge the luminous Sufi Masters—their startling actions, their Sufi poetry, their revelatory tales.

5
Islamic Meditations
Oral Traditions of the Prophet Muhammad

THE FIVE PILLARS OF ISLAM

The Mercy of Allah to all Worlds reveals in his sublime Oral Tradition the Divine Attributes of Beauty and Clarity, reflected as the five fundamental principles of Islam.

The first principle, which contains the others in subtle form, is *la ilaha illallah muhammad rasulallah*. This movement of the soul begins with the spontaneous affirmation of Divine Unity. There is nothing outside Divine Unity. All creation exists only from and within Divine Unity. From this affirmation with our whole being of all-encompassing unity springs spontaneous conviction that perfect humanity, the crown of creation, is manifest through the Pearl of the Universe, the beloved Muhammad, upon him be peace, Messenger and mystic slave of Supreme Love.

The second root principle of Islam is to lead an entirely prayerful existence in continuous remembrance of Allah, spiritually perceiving all events flowing from the Will of Allah and therefore unshakably established in the state of praising Allah. This state of being is beautifully reflected in the five times of formal prayer. At dawn, noon, afternoon, sunset, and night, to pray with selfless love the graceful prayers that the holy Prophet received during his mystical ascension is to ascend with him to mystic union with Reality, according to his promise and the Promise of Allah.

The third guiding principle of Islam is the instinctive attitude of loving generosity toward all humanity, which flows spontaneously from gratitude to Allah for every breath of life. This state of being—exemplified by the noble Prophet, who was as uncontrollably generous as the wild desert wind—is reflected in the formal gift of one fortieth of our abundance each year to those in need. This spiritual station of generosity manifests through every thought and action, intention and gesture of the person who is surrendered to the All-Generous, All-Merciful, All-Forgiving, All-Sustaining One.

The fourth leading principle of Islam is to undertake the Great Pilgrimage to the Six-Dimensional Mirror, the radiant Cube of the Completeness of Revelation, the directionless Direction of Prayer, the Kaaba of Allah, the Holy House built by Abraham, the Center of the Universe, the True Heart of Humanity. While sojourning in the planetary Mecca and Medina—fulfilling all the subtle procedures of the pilgrimage, every detail of which is replete with spiritual meaning and power—one should perceive with eyes of the heart the heavenly Mecca and Medina, filled with true lovers, fragrant with selfless love. These souls are the very springs, fruits, and human pearls of Paradise. Through this perception, one abides in the constant station of pilgrimage.

The fifth fundamental principle of Islam is to respond wholeheartedly to the ecstasy of Ramadan. Radiant joy is poured by Allah into the entire universe from the open gate of His Paradise during this holy month to commemorate the descent of the Transcendent Quran into the adamantine heart of His Prophet. Our formal response to Ramadan is to fast from food and drink during one lunar month from before dawn until sunset. Our inward response to the holy Ramadan is to become so lost in remembrance of Allah that we do not feel the need for any power other than Allah, any sustenance other than the Love of Allah, or any distraction from the boundless Light of Allah, Who is the Light of the heavens and the earth. Ramadan enables us to experience living only in Divine Light, only through Divine Light. During the holy month, the Light of Revelation uncompromisingly reveals our gross and subtle selfishness, then mercifully dissolves it. Ramadan is not just a month-long practice but a grounding principle of Islam. Similarly, the affirmation of faith, the prayer, the charity, and the pilgrimage are not just particular events but underlying principles, continuous and all-encompassing. They are not merely passing states of consciousness but established stations of being.

The Messenger of Allah makes clear and vibrant in the heart of humanity the nobility of life natural to the human soul from eternity that is called Islam, teaching:

> *Islam is based on five—to testify that nothing can be worshiped but Divine Reality and that Muhammad is the Messenger of Allah, to offer the prayers carefully, to give in regular charity, to perform pilgrimage, and to fast during the month of Ramadan.*

LOVE FOR THE PROPHET OF ALLAH

No person can experience the complete, spontaneous affirmation of Reality—not only with every prayer or even with every breath but with one's very life and being—before loving the fully awakened human being more than any other, including parents, children, and all humanity. Love for the Messenger is the way to awaken to the crown of creation, to the true humanity within all persons. To love Muhammad intensely—even above the mystic shaykhs who inherit the spiritual wealth and beauty of the Prophet—lifts one into perfection. This state of completeness is simply to be through the Being of Allah, to live through the Life of Allah, to see through the All-Seeing, to hear through the All-Hearing, to know through the All-Knowing. This is the true affirmation of Unity, which is not merely verbal, conceptual, or conventionally religious. This is Allah affirming Allah through perfect humanity. This is *la ilaha illallah muhammad rasulallah.*

To love any human being, no matter how close or dear, who has not reached perfection more than one loves the perfect human being places further veils over the light of the hidden perfection of one's own humanity and the humanity of the other. To love the perfect human being above all other human beings, by contrast, removes obscuring veils from the hidden perfection of one's own being and the others whom one loves. To love the Prophet Muhammad, who is the Lover of Humanity and the Pearl of the Universe, vastly accelerates our progress—our awakening to ourselves and to others as pure souls, as rays from the Source of Light, who are simply the Affirmation of Unity blossoming as the Divine Drama of Revelation. Loving the noble Muhammad of Love enhances immeasurably our love for other human beings, on all levels of development, by revealing the infinite reservoir of love within us. Failing to love the Muhammad of Love above all beloved persons saps the energy of our love, making it finite and linking it with imperfection.

Love him for his perfection of love! Love him as the light of true humanity within every heart! We place our hand upon our heart when we speak his most beautiful name, Muhammad, because he is the true human heart—the essence of our own mystical heart, hidden far behind the heart of life throbbing within the human breast. He is not in the desert of Arabia, but in the heart. Love him, and you will love all persons abundantly. Fail to love him, and you will fail to love anyone truly. Love him who is the perfection of human love, and you will overflow with love for Allah, Who is the Source of Love. Disappear with ecstatic love into the Prophet of Allah, and you will disappear into Allah in mystic union. Fail to lose yourself in the love of the Prophet, and you will never directly encounter the Love of Allah.

Walk only on the way of Love. Love only for the sake of Love. Live only for the Good Pleasure of Allah, the Essence of Love. The Caretaker of the Ocean of Love, the uniquely beloved one of Allah, is the very principle of Allah's Self-Revealing Love. To love Muhammad, the Most Praiseworthy, is the unveiling of Divine

Love. Love unveiled is the play of lover and Beloved, the very life of the universe, the song of atoms, the sweetness of the human touch, the fragrance of prayers. The rose perfume of the Prophet's love drenches his lovers.

Thus Muhammad Mustafa, the Mercy of Allah to all Worlds, calls out through his living Oral Tradition:

None of you will have faith, till he loves me more than his father, his children, and all humankind.

HUMANITY

Allah has placed His Essence secretly in the heart of humanity. This is why the All-Powerful One commanded radiant ranks of angels to bow before the transcendental Adam, whom He taught the names of all beings and the Beautiful Names for the Source of Being. Every soul responded *Yes* to Allah Most Sublime when He tenderly invited us in the preeternal Universe of Souls to return to Him freely along the Path of Love. Every soul, no matter how heavily obscured by the veils of the limited self, is an expression of that original assent to Allah, that primordial *Yes,* that spontaneous submission to Allah and longing to return completely into Allah. Therefore, every soul can authentically receive in its holy essence the prostration of the angels.

The innate spiritual beauty of every soul shines forth in the beauty of the Prophet of Allah, who teaches and demonstrates the nobility of the human soul—its uprightness, its dignity, and the Divine Light that shines through it by its very nature. Islam is the teaching, the actual demonstration, of Divine Unity and human dignity. All-Embracing Unity and awakened humanity are inseperable. The Holy Quran reveals that Allah created humanity as the crown of creation, and the Prophet of Allah invites us to wear the crown of humanity. Worship of Allah only, perception of the radiance of Allah wherever one's gaze wanders in the vastness of Allah's perfect creation—this is the principle of humanity, the principle of universal Islam.

Who perceives Truth? Who loves Truth? Who lives Truth? Human beings do. The human heart alone consciously affirms *la ilaha illallah.* Actual, vibrant, clarified, beautiful, sensitive human beings—mirrors of the Divine Attributes—love and live the Truth of Islam. Islam exists for humanity. Islam is humanity. The revelation of Divine Books through chosen Prophets comes into being solely for humanity. The universe as a temporal training ground for timeless souls comes into being solely for humanity. And humanity exists only to reveal Allah, the Hidden Treasure. There is no fundamental purpose for the existence of human beings other than to affirm the existence of Allah with freedom, dignity, and delight. Through human existence, Allah affirms His Own Reality. The angels are created to protect and guide humanity. The earth, with its subtle balance of life, is created to support humanity, and humanity is appointed by Allah to function as the sensitive caretaker of planetary life. Human beings do not exist for some finite reason. Human beings exist only for Allah. How can we conceive the exaltedness of the human station? The name of humanity is inscribed with the Most Holy Name of Allah from before eternity as *la ilaha illallah muhammad rasulallah.* The whole universe, created for humanity, is contained subtly in the microcosm of the human body. Allah Most Majestic and Resplendent, who cannot fit into the entire cosmos, can enter and abide in the human heart.

Islam teaches and demonstrates companionship among human beings who live in the Peace of Allah, persons who lovingly offer each other the highest gift, which is the Peace of Allah. The holy greeting *as-salam alaykum,* may Divine Peace be with you, represents the most sublime fruit of the practice of Islam. The heart must actually contain the Peace of Unity in order truly to offer the Peace of Unity, which is Islam. When one genuinely gives the greeting of Allah's Peace, one is led spontaneously to give the nourishment of Allah's Love to human beings—in the form of food, help, livelihood, compassion, knowledge, and spiritual teaching. When operating from the level of Divine Peace, we lay down our individual, separate lives for others. The dervish lover of humanity refuses to enter Paradise until all human beings have safely entered. We stand together for prayer with humanity, inviting all humanity to Truth.

When human beings have not debased their humanity, human consciousness is the most powerful channel for the Love and Abundance of Allah. Through the hands of our brothers and sisters, we receive the blessings of Allah. Through these blessed hands we receive food to break our fast in Ramadan. Through generous hands we receive our precious copy of the Holy Quran. Through friendly hands we are helped to make the Great Pilgrimage. By tender hands we are brought into the world as infants and cared for throughout life. By reverent hands we are laid in the mystical marriage chamber of the tomb.

O sweet humanity, known and unknown! The perfect human being, the Culmination of Revelation, was asked, "What are the truly good, the supremely blessed, the most profound actions and attributes and qualities of Islam?" The Mirror of Allah's Love for humankind, the beloved Muhammad, replied with great simplicity in his incomparably gentle tone:

To feed and greet those whom you know and those whom you do not know.

PRAYER OF ISLAM

The Prayer of Islam was first revealed to the Leader of the Prayers of Humanity by Archangel Gabriel on a hillside above Mecca the Sublime. The most radiant angelic emissary of Allah struck the ground with his miraculous heel and a fountain of sweet water sprang forth from barren stone. He then showed the Prophet of Islam how to take ablutions, so that lovers of Allah could purify their entire physical, mental, and emotional being before experiencing the ascension of the Prayer, which lifts one's whole nature into the secret sanctuary of Allah Most High without leaving any dimension of true human reality behind. The holy Prophet, may the Peace of Allah always stream through him to all the worlds, then returned home and made the first Prayer of Islam with the noble Mother of the Faithful, the one who first recognized the spiritual leadership of her husband, the exalted Khadija, may Allah always illumine her soul.

Later, when the lover of humankind ascended on his mystical Night Journey to intercede at the Abode of Essence, the full meaning of the Prayer as the mystic ascension of the one who performs the Prayer was revealed. The beloved Muhammad then received Divine Assurance that five formal prayers a day would carry the transformational power of fifty prayers a day, elevating everyone who prays thus into constant remembrance of Allah, the exalted station attainable before only by ecstatic saints and ascetic sages.

The Prayer of Islam is given to lift all human beings without exception into the highest levels of mystical experience. No one who longs to ascend is to be left behind. This is the culmination of Revelation—the distribution of the Light of Prophecy that has shone on the foreheads of 124,000 Prophets, upon them all be peace, to the heart of everyone who stands for the Prayer, facing Truth alone, making the prostration of pure submission and calling out the Words of Transcendence, *allahu akbar.* This primordial Light of Prophecy, the *nur muhammad,* now streams through the Prayer of Islam and shines from the foreheads of all who plunge into the refreshing and reviving stream of the Prayer five times each day—standing, bowing, prostrating, and kneeling in purified bodies that anticipate the radiant bodies of the Resurrection. The whole earth has become a holy mosque. Continuous waves of the Prayer spread around the planet as dawn and dusk follow one another rhythmically, gracefully guided by the Source of Harmony, punctuated by melodious human voices singing the Call to Prayer from countless minarets of exultation.

The Glorious Quran is written in the very structure of the cosmos as well as in the letters of the Arabic alphabet. The delicate light before sunrise is Dawn Prayer, the subtle twilight Evening Prayer. Birds pray by opening their wings for flight, as the Holy Quran reveals. The entire creation is making the graceful movements of the Prayer. Galaxies are in full prostration. This green earth has been spread out for us as a rich prayer carpet. The sacred Valley of Revelation, Mecca the Ennobled, is the directionless Direction of Prayer. Angels gather at the various times of prayer, as Quran discloses, to enjoy sweet Arabic chanting and to observe beautiful human hearts in prostration and ascension.

The sublime interlude of Ramadan, the mystic lunar month when ordinary time becomes transparent to the timeless radiance of Paradise, is the Month of Prayer. The twenty-eight days of Ramadan are prayer, whether one sleeps or stays awake. And each night is a journey into the interior riches of the treasury of Allah's Beautiful Divine Names, in loving community and communion with brothers and sisters in the Peace of Allah. Every thought and action in Ramadan can become prayer, if our ablutions are powerful and constant, inwardly and outwardly. To fast on Ramadan is to take ablution from the very spring opened by the angel Gabriel for the Prophet of Allah on the mountainside. Prepared by fasting from before dawn, refreshed by breaking the fast after sunset, we enter the night prayers of Ramadan, which extend into eternity. On each holy night of Ramadan, time and eternity meet. The lovers lose themselves in prostration, surrounded by the fragrance of Paradise and hearing the music of the Resonant Quran as it resounds in Paradise, emanating from flowing rivers of ecstasy and sighing like wind in the branches of the fruit trees of wisdom.

These sacred nights, brighter than day, contain secretly among them the Night of Power, the diamond at the heart of the Treasury of Love, the Night into which is condensed the power of a thousand months of daily prayer. The fast of Ramadan is an intensified ablution for the prayer of this Night of Power. On this hidden night during Ramadan, time and eternity not only touch but merge completely. On this night, the Peace of Allah descends in fullness into the hearts and minds of all beings, even into the atoms of the physical creation. For this Night above all nights, we long to keep awake. To prepare for this revelatory Night, we pray the extensive prayers of Ramadan. Every cycle of prostration is a rung on the ladder of light leading to this majestic Night. On the Night of Power, the Glorious Quran descended into the personal awareness of the Prophet from the heavenly awareness of his soul, which was immersed in the Holy Quran from eternity. On the anniversary of this Night of Power, the Prophet ascended on his Mystic Night Journey along the very way that the Holy Quran had descended. Every Ramadan on the Night of Power, the transcendent Peace of Allah descends along the same mysterious route. On every Night of Power, the lovers ascend this very stairway of light to mystical union with the essence of Reality.

The entire Ramadan is the Night of Power. The gem of this supreme Night has twenty-eight main facets, each appearing as a shining night of prayer while the diamond of the sacred month slowly turns. Each main facet shines with the energy of different Divine Names and is adorned with twenty smaller facets, which are the sparkling cycles of the extended prayers offered on each night. Through this Diamond of Clarity shines the Light of Forgiveness, bathing our souls and bodies in its splendor, healing us in every way, conferring on us true sincerity—the knowledge of Divine Unity—aligning us with the Good Pleasure of Allah, and clearing from our minds the play of mere opinions and mundane concerns.

Between his desert dwelling of humility and his pulpit of the glorious teaching of unity, the praiseworthy and victorious one, the beloved Muhammad Mustafa, upon him be peace, stands eternally within the Great Mosque of the Medina of the Heart, leading the night prayers of Ramadan. His message from the boundless Mercy of Allah is unequivocal:

> *For whoever establishes prayers during the nights of Ramadan, out of sincere faith and hoping only to attain Allah's Good Pleasure, all past errors and negations will be forgiven.*

TOLERANT WAY OF TRUTH

The beloved Abraham, whose station is to be the intimate friend of Truth, is a Prophet whose eternal presence, subsisting with and within Allah, guides the mystical return of humanity into the transcendent Heart of Reality. The special responsibility of the noble Abraham is *haqiqa,* the realization of Truth alone. Here the path of religious duty, the *sharia,* and the path of contemplative practice, the *tariqa,* both enter the pathless field of supernal radiance, plunging like rivers of love into the Ocean of Love. The magnificent patriarch Abraham, the embodiment of true generosity and hospitality whose tent is open in all directions, is spiritual father to the Jewish, Christian, and Islamic nations, which are his seed multiplied into galaxies of souls, according to the Divine Promise.

Embodying the fundamental religion of humanity, the religion of unity, the Prophet Abraham, upon him be abundant and beautiful peace, is the *hanifa,* the upright one. Even before the Divine Revelation of Torah, Psalms, Gospels, and Quran, he demonstrates the original nobility of the human being in direct communion with Truth, without images or mediators. As the Light of Prophecy intensified, approaching its culmination, through this most intimate friend, Allah revealed the meditation on Ultimate Unity. This contemplation transcends the entire spectrum of Divine Attributes. The Glorious Quran discloses that while he was deep in prayer and meditation during solitary desert retreat, the noble Abraham received the direct experience of Essence, alone and perfect, needing no creation and no Creator. This Essence abides beyond the Attributes of Allah—Star of Majesty, Moon of Beauty, Sun of Power—which are mirrored as creation. The sole existence of this formless, creationless, attributeless Essence as the only Reality is the mystic teaching that beloved Abraham brought to humanity as a further evolution of the understanding of One God, which has been repeatedly revealed by Allah since the Prophet Adam, upon him be peace.

The beloved Muhammad, the final fruition of the evolution of humanity's understanding, the final link in the lineage of revelatory Light, calls the spirituality innate to the human soul the *hanifiyya,* the way of Abraham. This station of maturity is wisely tolerant of all levels and dimensions of religious understanding, manifest through Prophets who, as Quran discloses, have been sent as Divine Signs to every nation without exception, demonstrating a clear path to the One Reality. This loving tolerance of *hanifiyya* takes as its symbol the Tent of Abraham, open to spiritual travelers from all directions, full of the abundant Generosity of Allah poured out to all seekers of truth without exception. The holy lovers on the way of *hanifiyya* inherit the original friendship of Prophet Abraham with living Truth. These lovers become friends as well to all expressions of Truth, to all precious human beings. Thus the *haqiqa,* the realization of Truth—the pure vein of gold inside the mountain of outward appearance, the translucent pearl hidden within the rough, dark oyster shell—accepts all human beings as who they truly are. The *hanifiyya* bestows inconceivable spiritual wealth—the golden radiance of Truth, the Pearl beyond Price—upon every human heart, from whichever direction it approaches the Tent of Abraham.

May humanity as a whole become conscious of the *hanifiyya,* allowing ineffable harmony to blossom among apparently diverse cultures, sciences, and great religious traditions. The perfect reconciler of the countless stars from the Universe of Souls who have become the children of Abraham, the Prophet Muhammad, upon him be peace, therefore declares:

The religion most beloved to Allah is the tolerant Way of Hanifiyya.

MERCY

The flood of goodness that enters the universe through the vast channel of universal Islam is immeasurable. The Mercy of Allah culminates as the noble Muhammad, upon him be peace. Through this all-awakening, all-transforming *rahman* and *rahim,* this inconceivable Divine Mercy and Compassion, the Prophet of Peace who is the Seal of Prophecy has become the complete Mercy of Allah to all humanity and to all worlds. The Messenger of Islam has come for those who have accepted the sublime invitation to the incomparable Divine Peace, to those who are making sincere efforts to open their whole hearts to the peace that can spring only from the perfect affirmation of Unity. He has come as well for those who, without knowing, are part of the spiritual body of universal Islam by virtue of their noble human souls, which are by original nature oriented toward Allah and freely returning into Allah.

The uniquely beloved one of Allah, may supernal peace surround him and stream through him to all humanity, speaks spontaneously from his station of constant Divine Expression called *marifa.* These precious words are written upon the hearts of his companions—upon careful pages of love with the black pen of awe and the green and golden ink of beauty. These utterances become the great, flowering forest of *hadith,* the Oral Tradition, as full of spiritual secrets as the Glorious Quran. Both emanate from precisely the same Source of Wisdom, the Only Source, the Hidden Treasure, ever-revealing yet hidden still. The inconceivable, or hidden, nature of Allah, who is more than Sublime, more than Most High, is evoked ceaselessly by the prayerful cry of His lovers, *allahu akbar,* Allah is always greater. The Prophet of Mercy, acting in mystic union with Divine Mercy, reveals the hidden Mercy of Allah in his illumined Oral Tradition, speaking about the unspeakable and giving spiritual measure to the immeasurable.

Through these most sweet, bearded lips, the sacred mathematics of Allah's Mercy issue like a stream of light, revealing that the response of Allah to a single intention of goodness is at minimum tenfold and, depending on the depth and purity of the intention, as much as seven hundredfold. Each sincere intention to affirm Unity through prayer or through offering kindness to the creations of Allah draws forth this intensifying Divine Response. Good thoughts and actions, therefore, carry at least ten times the power of their own intrinsic goodness in the field of Divine Power called Islam. By contrast, intentions or actions of negation are never increased in intensity by Allah and can be easily swept away by the Divine Forgiveness whose power is absolute.

Among thousands of intentions and actions that stream forth every day from each human being—including each breath, each step, each thought, each perception—every negation remains single, if Allah allows it to remain at all, whereas every affirmation increases in power by at least a factor of ten. These are the moral and spiritual mathematics of the unified field of Islam. In a rich and long human life, full of the impulses of compassion and worship, the implications of this Divine Mathematics of Mercy are overwhelming. A single precious month of Ramadan could contain, for the pure practitioner, the spiritual fruitfulness of seven hundred Ramadans with seven hundred Nights of Power, each of which contains blessedness equal to a thousand months of prayer. A single meal offered lovingly and for the love of Allah to a hungry person could generate the goodness of offering seven hundred such meals. The deep gratitude to Allah of the recipient of this selfless kindness could be increased seven hundred times as well in its heart-transforming, world-transforming power.

For someone who has not been blessed by living consciously since birth within the Realm of Mercy called Islam or for someone who has fallen away and renews his or her witness to the Truth, the Prophet of Islam brings news of inconceivable joy. Allah Most High cancels all the previous negations engaged in by the ones who sincerely embrace or re-embrace Islam. These blessed persons begin fresh, with only the power of their previous positive intentions and actions remaining in the stream of their subtle being. This is the flood of goodness that enters creation through Islam and that increases geometrically and exponentially as Islam spreads consciously among humanity.

One must contemplate the surprising implications of this Mathematics of Mercy with the eyes of the intellect illumined and the eyes of the heart fully open to the trustworthy nature of the direct spoken words of *al-amin,* the absolutely trustworthy one, may the peace that he proclaims to the world always pervade his radiant soul. With deep spiritual understanding, may we hear the sweet tone of his voice that once spoke in the desert of Arabia and now speaks timelessly within the heart of humanity:

> *If a person embraces Islam sincerely, then Allah will forgive all his past errors. After that starts the settlement of accounts: the reward of his good actions will be ten times to seven hundred times for each good deed and any negative action will be recorded as it is, unless Allah forgives it.*

DEATH

Death can be the blessed transition from life lived by a limited individual for his individual aims, confined within his individual world, to life lived as a being merged in the boundless Light of Allah, for the boundless aim of Love, released into the boundless realm of heavenly awareness. The lovers of Truth, who die before they die—as the supreme mystic guide teaches us to do—make this transition fully while still living on the sanctified earth, created for humanity as a place of refreshment, as a prayer carpet and a planetary mosque, as a university for the training of souls. Thus the true companions of the holy Prophet who, by the Permission of Allah, awake and arise during every generation are dwelling, while they live on earth, in one of the ninety-nine levels of Paradise, or even in the Garden of Essence above Paradise.

Whoever loves these companions of Love, who have died completely to the limited self, and whoever longs to live their holy way of life will actually glimpse Paradise here on earth and will enter Paradise fully after plunging through the sleep of death. Whoever is indifferent to, or even opposes, the friends of Allah and whoever acts directly counter to their holy way of life will glimpse hell during this earthly existence and will enter the self-generated nightmare of hellfire after falling into the sleep of death.

Even those who enter the grave burning with the fever of their diseased conscience—twisting and turning in the agony inflicted upon their subtle being by the negativity of their own psyche now released from the restraints of the physical world—may receive Divine Grace. They will be gracefully removed from this torment by the All-Merciful One if their hearts contain even a tiny seed of love for Allah and the Prophet of Allah and his noble family, if their hearts contain even a pinpoint ray of the light that is the affirmation of Unity and is their own basic humanity. Utterly chastened, purged by psychic fire from the dross of their own chronic cruelty, indifference, and arrogance, their subtle bodies stinging with self-inflicted suffering, these souls will be plunged by Allah All-Merciful into the River of Life, the eternal flow of the beautiful Divine Name *ya hayy,* the All-Living One. They will revive, awaken, and enter the resurrection of Paradise, praising the All-Merciful with poignant intensity. Their bodies of radiance will be tinged by this terrible ordeal, like crystals with flaws, green plants with slightly yellowed leaves, or trees that have not grown straight.

The Mercy of Allah to all Worlds, the Intercessor for all Spiritual Nations, the Overseer of the Tumult of the Resurrection, the beloved Muhammad, may he be immersed in the Peace of Allah, teaches about this inconceivable Mercy of the All-Merciful One:

> *When the people of hell enter hellfire, Allah will order those who have faith equal to the weight of a grain of mustard seed to be removed from the fire. They will be taken out already seared.*

Placed into the River of Hayat, they will revive like grain that grows near the bank of a flood channel and that comes out yellow and slightly bent.

SPIRITUAL PERFECTION

Spiritual perfection, *ihsan,* is awakening to the ultimate nearness between human soul and Divine Reality. In this inspired evolution, we do not approach nearer and nearer to Truth because, as the Glorious Quran reveals, Allah is already closer to us than our own life. No one can move even the slightest distance away from Allah Most Near. Nor do we, through prayerful efforts, turn our own minds and hearts toward Divine Light, but, as the Clear Quran discloses, Allah alone turns us. Paradise is the beatific vision of Allah's true nearness to the soul—the blessed vision of Light that the friends of Allah receive fully while living on earth when, through Divine Permission, the surface of time becomes transparent to the depths of eternity. The experience of Paradise, here or hereafter, is not the automatic result of religious practices but is purely the Gift of Allah to the soul. No religious practitioner, however careful and devout, can be assured of a place in Paradise until Allah reveals the formless brilliance of the Palace of Vision that awaits the soul in the Garden of Nearness. Even then, the visionary cannot become complacent for a moment. The subtle hypocrisy of self-congratulation can obscure the diamond palaces and radiant gardens of Divine Nearness, which constitutes spiritual perfection.

The person submitted only to Truth, who has been shown his or her station in Paradise even while living on earth and who has thus entered the ever-expanding, ever-ascending dimension of spiritual perfection, must be more vigilant than ever. This person must be extremely careful never to ascribe any power to the limited self—particularly to the spiritual efforts of the higher levels of the self—but to ascribe all power and perfection to the Source of Power and Perfection alone. To maintain the existence of any separate self would be the fundamental illusion of placing partners beside Allah. Nearness alone gives, Nearness alone receives, and Nearness alone sees.

The sublime Ali, may Allah always exalt his soul in Nearness, proclaimed boldly, out of the ecstasy of spiritual perfection, that he would never worship any divinity whom he could not see. This was not his own teaching, for this submitted lover of Allah and intimate spiritual son of the Prophet of Allah would never claim any insight or even any breath or movement as his own. This bold affirmation of true knowledge was the noble Ali's confirmation of the words of the President of the Parliament of Prophets, the beloved Muhammad, may the mystic union with Divine Love be granted always to his sublime soul, to his beautiful family, and to his spiritual companions throughout time and eternity. The holy Prophet disclosed that lovers in the highest realm of Paradise will see Allah Most High with more clarity than we on earth perceive the full moon in the midnight sky. May Allah protect us from misunderstanding this as the vision of form, for Divine Reality is absolutely formless, boundless, indescribable, and inconceivable.

These trustworthy words of the trustworthy one that evoke the state of spiritual perfection, the state of seeing Allah as the brilliant moon in the empty sky of awareness, simply indicate the experiential tone of clarity, fullness, brightness, beauty, concentration, immediacy, and the absence of multiplicity. Even the brightest stars disappear when the full moon rises. Thus even the angels disappear during the beatific vision of Divine Light, as the Prophet of Allah confirms when he states that no angels intervened when his courageous and humble spiritual gaze was directed straight into Supreme Reality.

So Ali the Exalted was simply transmitting the teaching of his Shaykh, the Mystic Guide of Humanity. As a boy of thirteen, Ali powerfully affirmed the Prophethood of Muhammad and was designated by the Messenger, at that moment, to be his inward spiritual successor. This noble warrior, called the whirling Lion of Allah, saved the precious life of the Prophet three times during righteous battle. Hazrati Ali is chosen by Allah and by the Prophet of Allah as the protector of the mystical path that ascends through countless radiant levels and degrees of submission and union. Ali is the root of the ever-flowering Tree of Tariqa, which brings forth fruit in all seasons of human history, in whose fragrant shade the dervishes have found and will always find refuge and delight. Thus all secret Sufis, ecstatic lovers of Truth, and poor *faqirs* who have lost their possessions and their personal awareness in Allah are under the special spiritual protection of the Great Ali, the foremost representative of the holy Prophet to his mystical community, the People of Nearness. The boldness of Ali must become ours. We must never worship out of a sense of conventional religious duty. The marriage of our soul to the Beloved will not be a social arrangement but an all-consuming passion of the heart.

We will not worship any Divine Reality that is distant, but will bow and prostrate only within Nearness, surrounded on all sides, above, and below by the *nur* of Allah, the Divine Light of the All-Seeing and All-Revealing One. We will clearly contemplate the mysterious full moon of unity with the enlightened eyes of the heart, which perceive in perfect immediacy, without distance, separation, or multiplicity. This is the spiritual perfection that has no ceiling or limit but calls the soul to race eternally deeper into ineffable Divine Light, drinking from the central fountain of Paradise, whose mystic name, revealed by the Glorious Quran, is *Ever-Seeking*.

The holy Prophet, may he be intimately embraced by Divine Peace, recounts that from one prayer period to the next he was always taken deeper into the knowledge of the inexhaustible Source of Knowledge. Spiritual perfection is never static. Eternity is an infinite spiral ascending into Nearness. The Seal of Prophets always speaks from his station as protector and revealer of the dimension of *marifa,* the Nearness that, in the radiant words of Quran, is *nearer than near and even nearer than that.* The

Bearer of the Words of Allah to Humanity can speak in simple language about Reality, which is absolutely beyond human language, because he remains in constant conscious union with Reality, expressing Reality through the channel of his beautiful presence. When asked by none less than Archangel Gabriel about the way to spiritual perfection, *ihsan,* the Intercessor for the Human, Subtle, and Angelic Kingdoms replied spontaneously and with sweet simplicity:

> *Ihsan is to worship Allah as if you see Him, and if you cannot achieve this state, to experience that He is seeing you.*

GIVING

The primary Attributes of Allah are *rahman,* supremely intensive compassion, and *rahim,* a mercy so panoramic, so total, so beyond the realm of calculation as to be entirely inconceivable to the human mind. All other Divine Attributes, from awesome Power and Justice to tender Beauty and Holiness, are subtly contained in *rahman* and *rahim.* Allah is the fundamentally generous Reality Who overflows spontaneously with the most tender mercy and Whose ontological nature is to give mercifully. Yet mercy cannot function fully without someone to receive and appreciate that mercy. Before time and eternity, Allah was a Hidden Treasure that desired to be known, to confer Itself, to give. This is revealed by Allah through the Prophet of Allah in the sublime Oral Tradition. The Hidden Treasure was Mercy, and creation came forth spontaneously to receive and appreciate that Mercy and therefore to praise that Mercy. The Self-revelation of Allah is the Self-giving of Allah. Mercy must give. Mercy can only give. So the perfect creations of Allah, the human souls that bear His Essence and reflect His Attributes, must become spontaneously self-giving, must become instinctively merciful, in order to be who they truly are.

Allah gives the richness of His Mercy to creation, and creation gives the ecstatic flood of its praise to Allah. All that is living is giving. And this primordial giving, which is the source, way, and goal of our being, is so total, so generous, that absolutely nothing is held back. The one who truly affirms *la ilaha illallah* gives so completely that no part of the personal being is retained. Whatever is held back from this prayerful self-giving becomes an idol. The immature religious person—one who does not give himself completely to Allah or who does not give to others purely for the sake of Allah—is unconsciously an idolater. His limited self becomes his idol. To give away the limited self entirely is true worship. To give others what they need as an expression of the worship of Allah is true giving, which is simply a channel for Allah's Generosity.

All the spiritual secrets of Islam are hidden in generous giving. Nothing should be taken for oneself. One should not even take a single breath. Instead, offer each breath to Allah. Nothing should be given to others for oneself—for one's own advantage, for one's own subtle enjoyment. Even a simple greeting should be given for the sake of Allah. One's very life should be given entirely to others for the sake of Allah. To give without ceasing, to give with total abandon, is to give as Allah gives. Divine Reality is All-Giving. The human reality must also become all-giving. Allah responds to selfless human generosity with this overwhelming Divine Generosity, which offers the soul Paradise, more precious than the eighteen thousand universes. Divine Generosity extends beyond this gift of Paradise. To the lover who gives the limited self away entirely, Allah responds with the inconceivable gift of His Own Attributes, adorning His selfless lovers with Divine Beauty, Divine Light, Divine Love, Divine Power. Beyond even the Attributes abides the Garden of Essence, the transparent Ground from which Divinity springs.

Allah's supreme gift, His Mercy to all the Worlds, the beloved Muhammad, may Allah bless him and give him peace, always insists upon the spiritual nature of every act of giving. Not only giving to the poor but even giving to one's own family, teaches the mirror of the Mercy of Allah, is a holy action that draws forth the response of Divine Generosity. Thus the holiness of giving touches and pervades even the most intimate moments of life in the human community, that is, the communion within one's immediate family. All gifts, no matter how personal or how small, can and must be given truly for the love of Allah, transforming every dimension of our existence into perfect giving. Our hands and heart then open entirely, like a flower spontaneously giving its fragrance. Our only prayer becomes: "Accept our intention to give, O Allah. Please accept it for Your sake only. Transform our life itself into giving."

As the most generous soul, the beloved Muhammad, so clearly teaches:

If a person spends on his own family sincerely for Allah's sake, then it is truly charitable giving. You will be richly rewarded for whatever you spend for Allah's sake, even for every morsel of food you put in your wife's mouth.

KNOWLEDGE

Knowledge is the jewel at the center of the setting of faith. Direct knowledge of the One Source of the Universe alone constitutes the total peace that is called submission or Islam. This true submission comes only from knowledge of the all-embracing Unity, which has no boundary, so that nothing can exist outside or beyond it. Clear knowledge of Unity has been revealed through all the authentic Holy Books transmitted to humanity through the Prophets from the single Source of Wisdom. Since knowledge about the creation is an extension of knowledge of the Creator, all knowledge is sacred. The fields of literature, science, medicine, jurisprudence, philosophy, religion, and the mystic path are not separate from each other. They are one harmony of knowledge.

The most precious gift of Allah to the human being is the sensitive intelligence, by which true knowledge is received. Human intelligence in its innate purity is the mirror of Divine Reality, Who is infinite Intelligence. Our clarity and balance, our ability to perceive accurately and to think validly, the brightness and calmness of our awareness—this is the living intelligence that the Creator, Exalted be He, has placed within all dimensions of the human person. This human intelligence is designed to become the vehicle for Divine Knowledge. We abstain from intoxicants simply to protect the innate purity, clarity, and balance of our intelligence. We do not allow ourselves to become fanatical or overwhelmed by passions simply to protect this finely tuned instrument, to keep free from pollution this pure stream of intelligence that flows through us from the Source of Intelligence.

This sacred intelligence is the Gift of Allah to our being through which we praise Him, love Him, stand in awe of Him, study and meditate upon His Divine Revelation, and follow His Divine Commandments. Through this scintillating intelligence—alive with the very Breath of Life, the All-Living One—we discern the interior path and receive the sublime teachings of the masters of the mystical Way of Return. This intelligence alone allows us to recognize with ecstatic certainty both the stages and the goal of the spiritual path. Through the door of this shining intelligence, we enter and disappear into infinite Intelligence, the Light that is Allah. Through this diamond intelligence, we contemplate and eventually become the Attributes of Allah, perceiving and understanding with Allah's Own Knowledge.

To the All-Knowing One, the Best and Only Knower, *ya alim,* all creation is transparent. This very transparency is known to the person of knowledge. The entire organic universe and the hierarchy of spiritual realms constitute the Universal Quran. These cosmic pages are turning spontaneously before our eyes of knowledge, inviting us to constant study, to vibrant wakefulness. The holy person of awakened intelligence learns from Allah during every waking moment and also through the deep knowledge revealed in dreams. There are as many levels of knowledge as there are individual letters in the Holy Quran. All of them are unified and complementary, as the letters of Quran are. The knowledge of Unity shines through every event and phenomenon in creation. The knowledge of Unity is the very principle of thinking, perception, language, and communication. The knowledge of Unity is the foundation of love. There is no opposition between love and knowledge. To love is to know, and to know is to love.

The ecstatic affirmation of Itself by Transcendent Unity through the lips of Mansur al-Hallaj, may Allah exalt his soul and guard his secret—the cry of *ana-l-haqq,* I am Truth—is nothing less than the union of supreme knowledge and supreme love. Love is not emotion or sentiment. Love is the knowledge of the names of all beings and the Names of the Source of Being taught by Allah to Adam, upon him be peace. The beloved Muhammad proclaims, "The person who knows himself knows his Lord." The person is knowledge and the Lord is Knowledge. Al-Hallaj and the realized sages in every culture, recognized or unrecognized, who have reached the goal of true self-knowledge and have thus become one with Divine Knowledge are the living confirmation of these words. All Prophets have taught the Path of Knowledge. "Know the Truth and the Truth will set you free," exclaims the beloved Jesus in the Holy Gospel revealed through him by Allah Most High.

The Culmination of Knowledge, the Distributor of Mystic Knowledge to All Hearts, the Perfect Man of Knowledge, the Rain of Knowledge and Intelligence upon All Creation, describes himself in this way: "The guidance and knowledge with which Allah has sent me is like abundant rain falling on the earth." May we walk humbly upon the way of knowledge, which Allah affirms in His Book of Clarity and Certainty as *la ilaha illallah* and demonstrates through the Oral Tradition, the enlightened intelligence of the Prophet Muhammad, may the perfect peace of knowledge always be his. The Messenger of Allah recounts:

While I was sleeping, I saw that a cup of milk was brought to me. I drank my fill, until noticing that milk was streaming out through my fingernails. The companions of love eagerly asked me about the true interpretation of this dream, and I responded simply, "It is knowledge."

LA ILAHA ILLALLAH

The power and holiness of Islam flow from the affirmation of Divine Unity, *la ilaha illallah,* the formless worship of Allah that removes the gaze of the heart from creation and concentrates spiritual vision upon the boundless expanse of conscious Unity, within which creation blossoms. Before the Divine Movement of Creation, before the angels circled the Throne, even before the manifestation of Divine Majesty called the Throne, Allah was the Hidden Treasure, Essence without attributes. The Treasure longs to be discovered. This Divine Longing is the preeternal lightning flash of the formless Muhammad of Light within the dark rain cloud of Essence. This flash of revelatory light, since it remains before time and eternity, is still at the moment of its original flashing forth, still fresh and new, the first dewdrop of light in the Garden of Essence. This original *la ilaha illallah* exists timelessly in the secret heart of humanity, which has never left the Heart of Divinity. This secluded inward place is where the dervish chants the sublime Word of Unity, which is the very Word of Power by which Allah creates and withdraws the universe. Here, in pristine inwardness, illuminated brilliantly by the continuous lightning flash *la ilaha illallah,* the rain cloud of Essence thunders sonorously with the Divine Response *muhammad rasulallah.*

The abundant rain of revelation now begins to fall. The Attributes of Allah emerge as eternity, containing the universe of all souls, who will eventually incarnate into time. Then the Divine Attributes reflect through temporality as solar systems. This inconceivably vast field of planetary life contains human civilization and Prophets bearing the Light of Prophecy—*la ilaha illallah,* the message and energy of the worship of Allah only. For what appears in time as ages, the thunderhead of Essence, opened by the First Light, the dynamic flash of *la ilaha illallah,* pours forth the rain of Divine Abundance, Divine Generosity, Divine Mercy, Divine Revelation.

The culmination of this storm of love is reached when the diamond soul of the Prophet Muhammad, may he remain merged in the peace of *la ilaha illallah,* descends to earth and the Radiant Quran descends through him as the complete revelation of the secrets of Allah. This Resonant Quran now encircles the entire planet as the sweet music of daily prayer. The original lightning flash, *la ilaha illallah,* the Muhammad of Light, has finally become the Quran-bearing, Truth-bearing Muhammad of Arabia, shining as the delicate moonlight of tender wisdom upon the landscape of the human heart. The First Light and the Final Prophet are one *la ilaha illallah,* one worship of Allah only. Between the First Light and the Final Prophet there is no distinction in essence, though the First is formless and the Last takes the beautiful form of the perfect man, the spiritual form that still appears to his lovers in dreams and visions.

The 124,000 Prophets who have streamed from the Heart of Unity are the one light of the one religion of the one worship of the one God by the one humanity in the one creation. The dazzling diversity that blossoms within Divine Unity, never disturbing this Unity, always clearly expresses oneness. There is only one. Allah and His infinite Creativity are one. The veils of creation are none other than the Power of Allah. They are transparent to Divine Light for the person who worships Allah only. And there is nothing other than the worship of Allah. Wherever the enlightened human gaze travels in the universe, there is the praise of Allah. Wherever the illuminated human soul voyages in the subtle heavenly realms, there is the praise of Allah. Allah alone is worthy of worship because Allah alone is. To be is Allah. To create is Allah. To worship is Allah. To praise is Allah. To love is Allah.

La ilaha illallah is the Self-affirmation of Allah, not some existence separate from Allah that is affirming Allah, for there is no reality apart from the One Reality. *La ilaha illallah* means that there is nothing other than Allah. Only Allah is Allah. Only Allah is. Allah is. As the Most High revealed through His Light of Prophecy to the noble form of the Prophet Moses, "There is nothing apart from the boundless *I Am* that I am." *La ilaha illallah* is Allah crying, "I alone am." This is the Truth that the Prophet Muhammad, filled with the nectar of the Divine Longing, longed to reveal to his beloved family and companions, and through them to all humanity—*la ilaha illallah*. There is nothing further to reveal. Beyond all-embracing Unity, perfectly expressed through humanity, there can be no further development. The lover of Truth who sings *la ilaha illallah* has seen the dawn of the Last Day, has been welcomed by the Voice of Allah, and has tasted the delight of the highest Paradise. The Hidden Treasure, the First Light, and the Last Day are one. Remaining outside time, they are immaculate and immediate in the heart. The Seal of Prophecy awakened the universal heart of humanity with the Divine Call *la ilaha illallah*. The Divine Movement from hiddenness to manifestation is now complete. The First Light affirms *la ilaha illallah*, and the Last Prophet affirms only this. *La ilaha illallah* is the mystic return of the universe into Allah. The Cosmic Quran is only *la ilaha illallah*. *La ilaha illallah*, which marked the first emergence of eternity, has now become the final Day of Resurrection.

La ilaha illallah is the compassionate intercessor for all souls. A cherished companion on the Path of Love once asked the holy Prophet, "Who will be the most blessed person and gain your intercession on the Day of Resurrection?" The Messenger of Allah replied with great tenderness:

> *I thought that no one would ask me this before you, as I know your longing to receive the words of my Oral Tradition. The most blessed person who will have my intercession on the Day of Resurrection will be the one who repeats sincerely from the depths of the heart, la ilaha illallah.*

TREE OF TARIQA

The *tariqa,* the steep path of ascension within Islam, manifests subtly as a living tree. Its vast limbs and smaller branches bear the honorable names of mystic shaykhs, the Friends of Allah who have branched forth as Divine Knowledge and Divine Love from the single trunk, flowing with the sweet, life-giving sap of the direct perception of Reality by Reality. Beautiful dervishes are the shining green leaves of this tree of union that shoots up in the secret garden of Unity, protected by white marble pillars, the fundamental principles of Islam. These living leaves tremble ecstatically in the wind of the Holy Spirit, absorbing the dew of the constant remembrance of Allah and the sunlight of Divine Essence. Leaves are not simply an adornment of the tree but the life of the tree. These innumerable leaves of dervish souls are one with the stems of the shaykhs from whom they are intimately springing. The stems, in turn, are one with the branches of the Grand Shaykhs, the founders of mystical Orders. Leaves, stems, and branches emerge in graceful patterns of Divine Will. The beloved Breath of Allah, the Messiah Jesus, may peace be upon him always and upon his Virgin Mother, describes his subtle relation to the disciples: "I am the vine and you are the branches."

The living root of this Tree of Tariqa, manifest within historical Islam, is the Sultan of Dervishes and the Guide of Mystic Guides, the beloved Muhammad, may Allah embrace him forever in the most intimate union. The fundamental name of this tree is *tariqa muhammadiyya.* There is only one *tariqa,* or mystical Order, in Islam—the universal community established by the Prophet Muhammad. The trunk of this miraculous tree is the sublime Ali, may Allah enlighten his spiritual countenance.

All knowledge contained by the Glorious Quran is subsumed in its opening chapter, Sura Fatiha. This vast knowledge is perfectly condensed into the initial words of that all-embracing Sura: *bismillah ir-rahman ir-rahim.* This consummate knowledge is subtly yet completely manifest in the first Arabic letter of the *bismillah* and is compressed, as coal is compressed into diamond, as the single point placed below the Arabic letter *b*. By the permission of Allah, the noble Ali unequivocally proclaimed, "I am that point beneath the *b* of the *bismillah*. I am the gate to the city of knowledge." The bearer of this exalted spiritual station was honored by the Prophet of Allah with the transmission of the outward *dhikrullah*, the ecstatic chanting in powerful tones of the beautiful Names of Allah, which Hazrati Ali would often practice alone in the wilderness—like a powerful desert storm, like the thunder of Allah's Own Storm of Love. This most exalted among the exalted companions of Divine Love is the awesome trunk of the Tree of Tariqa, which grows from the root of Muhammad and can also be called *tariqa aliya*. The truthful one, the most intimate companion in loving friendship with the Messenger of Allah, the first successor of the holy Prophet to lead the noble nation of Islam, Abu Bakr, may Allah always exalt his soul in Truth, received the transmission of the secret inward *dhikrullah*. This teaching occurred when the two friends were concealed together in the cave of Silence and Delight. The outward *dhikrullah* of Ali and the inward *dhikrullah* of Abu Bakr coexist in the subtle being of the dervish and in the life of the mystic way, reflecting infinitely like two clear mirrors facing each other.

The noble wives of the Prophet and his inseparable male companions in holy struggle, along with their wives, constituted the first dervish community of Islam, under the careful attention and guidance of the first and foremost Shaykh of Islam. The vast range of esoteric knowledge and practice among the mystical Orders of Islam is nothing more or less than the original, most intense, most fruitful companionship between the Lover and Guide of Humanity and his close dervish companions. Nothing has been added and nothing is missing. The spiritual riches of Quran and Oral Tradition have been further unfolded by inspiration over fourteen centuries and will continue to be unrolled like a beautiful tapestry or calligraphy until the End of Time. May Allah Most High prolong His Mercy upon future generations of lovers!

This Tree of Tariqa is more stately even than the trees of Paradise envisioned by the Prophet of Allah and disclosed in his powerful Oral Tradition as trees so huge that one could ride a fast camel for a hundred years without crossing the fragrant shade that each one casts. The four major limbs of *tariqa* are the supremely accomplished shaykhs and diamond souls Ahmad Rufai, Ahmad Badawi, Abdul Qadir Gaylani, and Ibrahim Dusuqi, may Allah continue to elevate them to the most exalted station of mystic union. The shaykhs and dervishes of this Tree of Tariqa are more numerous than the human mind can imagine, and each one is unique. In every generation of Islam, these sublime lovers arise as the ancient Prophets once arose, many of them remaining entirely hidden from human recognition. We name just a few of these inheritors of the Wealth of the Prophets, sending them our heartfelt salams, in order to invoke them all and thereby experience our own spiritual body extending outward through space and time as root, trunk, limbs, branches, and leaves of this living Tree of Transmission.

The accomplished souls of these venerable lovers of Truth and our own aspiring souls are not separate entities, dispersed throughout fourteen centuries or scattered across the planet and the heavenly realms. Both awakened and awakening souls grow organically as one Body of Love, one transcendent Tree of Life, which is not found in the East or in the West. Each true shaykh and true dervish is the whole tree. The fruit of this tree is pristine awareness. The shaykhs and dervishes do not taste fruit from the Tree of Life: they become this fruit. The flowers of ecstasy, which blossom in profusion on the Tree of Tariqa, are never gazed upon by shaykhs and dervishes. These flowers of many colors and fragrances are the spiritually realized gaze of the ones who are lost to the world and found in Allah. There can be no outside observer or experiencer of mystic union. The shaykh guides the dervish to the spiritual station where the illusion and subtle idolatry of twoness finally disappears. The shaykh leads sensitive lovers from the Divine Names to the One Who is Named, using gestures of being, not concepts, ordinary words, or external rules.

One such mysterious gesture of the original Shaykh of Islam, the Prophet Muhammad, is blessedly remembered and taught as part of his sublime Oral Tradition by one of his original dervish companions. This dervish lover, lost in his noble Shaykh and inseparable from his Shaykh, thirsted more than any other for the teachings flowing through the heart of this inconceivably powerful spiritual guide. The beautiful dervish soul, the absolutely devoted Abu Hurayra, may Allah continue to exalt him in companionship with the Beloved, remarks that he did no business in the market, nor harvested crops, but simply remained close to the holy hem of the robe of the Lover of Humanity.

Abu Hurayra once confessed to the wearer of the robe beneath which there is only Truth, "I have heard many narrations from you but I forget some of them." The root of the Tree of Tariqa replied, *Spread out your upper robe.* Abu Hurayra recounts:

> I did accordingly. Then he moved his hands, as if filling them with something, and emptied them into my *rida,* saying, *Take this cloth and wrap it around your body.* I did it. After that I never forgot anything.

THE MYSTERY OF ISLAM

The boundless esoteric knowledge transmitted directly from Allah Most High to His perfect servant was miraculously contained in ninety thousand concise Words of Power. Within the Abode of Essence, this inconceivable communication flowed from the Source of Divine Attributes to the clear mirror of Divine Attributes, the beloved Muhammad, upon him be supernal peace. The Caretaker of the Infinite Fields of Meaning of the Holy Quran, may Allah bless him with blissful union and give him the peace of perfect knowledge, has transmitted the Complete Revelation to humanity as an Arabic Quran of consummate clarity and subtlety.

The Quran calls itself a *hidden book,* for it secretly contains references to whatever events, significant and insignificant, have occurred or will occur until the End of Time. May Allah mercifully prolong the sojourn of humanity upon this beautiful oasis of the earthly plane! Friends of Allah can discover in the Clear Quran, which is to them a bride lifting her veil for the bridegroom, whatever spiritual teachings awakened sages of Islam will give in the future as well as whatever inspirations the lovers of Truth have received during the long evolution of humanity toward its spiritual culmination. We are now blessed to be living in the exalted Days of Completion. Contained in seed form within the intricate configurations of the Arabic Quran are the voluminous and startling writings of Muhyiddin Ibn Arabi and Mevlana Jelaluddin Rumi, may Allah reveal to their souls eternally the secrets of knowledge and love. There are as many levels of meaning in the Quran as there are individual Arabic letters in its ocean of miraculous verses, or *ayats,* each of which is a direct sign from Allah Most High. Each of these levels of meaning, in turn, contains countless illuminations, predictions, and applications. The ninety thousand transcendent Words of Light given to the Messenger on his ascension and the thousands of infinitely meaningful Arabic verses given to him on earth, as well as innumerable Oral Traditions, some still being revealed in dream and vision, represent the mysterious expanse of Islam.

How much of this can we hope to learn? How much of this can we conceive? The greatest holy practitioners abide only on the threshold of the vastness of Islam. Knowledgeable, prayerful persons teach the fundamental level of the sublime way of Islam to each successive generation of the faithful. The Five Pillars of the Sacred Law represent a teaching of clarity and a beneficent guidance that all human beings can receive, assimilate, and practice more and more deeply throughout their lives. This is the foundation of Islam, open equally to all. But the realms of Revelation extend far beyond this. Upon these Five Pillars rests a magnificent Palace of Vision.

The beautiful dervish of the Prophet, Abu Hurayra, may Allah keep him in the companionship his soul longs for, once made this awesome proclamation: "I have memorized two kinds of knowledge from the Messenger of Allah. I have propagated one of them to you. If I propagated the second kind, my throat would be cut." This beloved companion of the Master of the Worlds was not referring simply to human violence but to the serious spiritual consequences of teaching the higher levels of Islam, the mystery of Islam, to those who are not prepared by purity of heart, by the permission of the Pir, by the direct touch of the Shaykh, by the fire of selfless love, and by the secret inward call of Allah. The trunk of the tree of the mystical Orders of Islam, the noble Ali, may Allah continue to deepen his love for the Prophet and for the daughter of the Prophet, gives this warning to teachers of Islam: "You should speak to people according to their ability to understand." The ability spoken of here by the sublime inward guide of the Tariqa of Ali is not intellectual brilliance or scholarship but the subtle receptivity and accuracy of heart that receives the clarity of revelation as a flawless, dustless mirror reflects without distortion. The ecstasy of the one who—through the blessed words, actions, and being of an illuminated guide—glimpses the higher levels of Islam must never be emotional or sentimental. Any careful person can follow a clearly-marked road through the broad valley, but great ability is required to traverse a narrow path across mountain heights, where the mystic traveler must often leap through empty space, breathing rarefied atmosphere, dazzled by unimaginable vistas, awed by precipices falling away into emptiness.

One of these mountaineers of spiritual life, the cherished dervish companion Muad, would respond to the Prophet's words only with the ecstatic proclamation, "Here I am before you, most joyfully at your service, O Messenger of Allah." Once he received a glimpse of the higher teachings from the lips that caused the sun to tremble. The Liberator of Souls proclaimed, "There is no one who sincerely testifies *la ilaha illallah muhammad rasulallah* who will not be saved from hellfire by Allah Most High." The intimate companion became inebriated at this vision of Divine Mercy—that not a single person who loves Unity and the Messenger of Unity, no matter what he or she does or fails to do, will be excluded from Paradise. "Beloved Master," he cried out with great intensity, "should I inform the people about this so they may enjoy the good news?" The holy Prophet, who holds secrets more profound than anyone can imagine, replied, "No. I am concerned that they might depend upon this secret instruction only."

May our balance, discipline, and maturity be great enough, and our ability to understand deep enough, that we may receive, without danger of misunderstanding, from the living hearts of the representatives of the Final Messenger, even a glimpse into the mystery of Islam—the palaces of vision, the pearls of realization, the stairways of light, the rivers of ecstasy, the gardens of love, the fruits of wisdom, the banquet of companionship, the robes of beauty, the drink of union, and the higher and higher thrones of spiritual stations without end.

Glory to the Lord of the Worlds!
amin amin alhamdulillahi rabbi-l-alamin

EARTH IS A PLACE OF PRAYER

The Glorious Quran reveals the sacred nature of this earth. The Most High has created planetary life as an oasis along the journey of the precious human soul, as a place where timeless awareness can be tested by living beautifully in the realm of temporal responsibility. The Clear Quran teaches humanity to plunge into deep gratitude, to meditate ceaselessly upon the signs of Allah's Mercy: simple earthly abundance, rain and seasonal growth, sweet springs and fruit-bearing trees, fields of grain, green pasture lands, timber from the mountains. The Voice of Truth, speaking through the exalted state of prophetic consciousness, instructs the human being to be gratefully aware of Allah's most simple gifts: fire for cooking, water for drinking and for the purifying ablutions before prayer, clean earth for the ablutions before prayer when water is not available. Allah presents humanity the dome of atmosphere as a vast blue canopy, sun and moon as brilliant lamps, and constellations scattered throughout the night sky, inspiring us to contemplate the immensity of creation and the infinite power of the Creator. From Allah alone flows the graceful rhythm of night and day—sun providing warmth and light for the life of action and night providing darkness and coolness for the refreshment of sleep, adorned by the spiritual dreams Allah Most Subtle sends to guide the soul.

This blue and green tapestry of earth—with fertile soil and pathways and dwelling places prepared by Allah Most Merciful as a mother prepares a cradle—is anchored by primordial mountains, the stability of manifest Being, and is protected by natural principles of harmony emanating from Allah, such as the subtle barrier that separates salt water from fresh water. All creatures, who are living parables for Divine Power and Love, are provided with their daily sustenance in perfect mutual interdependence. Humanity is appointed by its Creator, exalted be He, as the sensitive, reverent caretaker for this balanced planetary life, enjoying the Abundance of Allah in constant gratitude, joy, prayer, and submission.

The awesome Voice of Truth, speaking through the Arabic Quran, reminds human beings to attribute every success to Allah alone, crying out spontaneous praises to Allah Most High when experiencing the exhilaration of riding a swift mount or sailing a sleek ship before the wind. Allah reminds human beings to perceive their own awareness and intelligence, as well as the continuous stream of prophetic teachers and revelatory scriptures, as irrefutable signs of His Love and Power. These are flawless demonstrations of Divine Presence for persons who see clearly, think deeply, meditate profoundly, and love selflessly. The entire creation is revealed as the Cosmic Quran. Each event and each moment is an *ayat,* a verse or sign, inscribed upon radiant pages of Light by the Pen of Universal Intellect.

The Throne of Allah rests upon the unifying stream of living energy behind the surface of manifest Being. This Divine Creation that sparkles around and within us is not mere matter, functioning through impersonal laws, but is the perfect mirror of Divine Attributes—alive completely with Divine Life, permeated by Divine Mercy, unfolding according to Divine Justice, no matter how imperfectly we may perceive the All-Embracing One. As the Glorious Quran reveals, no soul is wronged by even so much as the point of a date stone. The beloved Muhammad, may the Peace of Allah always surround him, was asked by his Lord to gaze into this Divine Creation and to discover if there exists the slightest disharmony, fragmentation, or injustice here. The Prophet's gaze, the most powerful gaze of any human being, returned, as Quran reveals, weary from its vast journey and dazzled by the diamond facets of perfection. Creation exists only within Divine Light, which is the one light of the heavens and the earth. The infinite variety of manifestation, the never-repeating stream of Divine Creativity that flows harmoniously within Divine Unity, never disturbs Divine Unity, which contains not only countless galaxies but seven subtle planes of being. There are races of subtle creatures called *jinn,* as well as mineral kingdoms, plant kingdoms, animal kingdoms, angelic kingdoms, and the Crown of Creation, the human kingdom. In the human being, Allah's Essence of Awareness and Energy of Manifestation have miraculously intermingled and mystically joined.

The boundless Power and Love of Allah are expressed perfectly through every atom, every star, every life in every kingdom on every plane of being. The human body—these very hands, lips, and heart, instruments of praise that will be resurrected as bodies of radiance on the Last Day—contains the whole creation in microcosm. This mysterious human form displays the ninety-nine beautiful Names of the Creator, the thousand and one Names, and the endless Divine Names. Five billion precious human beings now live on the planet, with billions before and, Allah willing, billions to come. They are created as perfect instruments of conscious praise. These most marvelous beings in creation are engaged in the Drama of Purification at every point on the surface of the planetary plane of life, struggling to drive out the forces of negation from their own hearts. By the mysterious Permission of Allah, insubstantial dark forces have entered His Creation through the original rebellion among highly placed subtle beings in the sub-angelic realms. The history of Prophecy is the constant victory of Divine Light. The whole globe of water and earth, green plants and atmosphere—this unified field of living kingdoms—has become a place of prayer in the course of this holy struggle against negation. The five periods of prayer now follow each other with the same beautiful rhythm as night and day, as seasons of the year and movements of the stars. The graceful rhythm of creation emerges from the Guidance of Allah alone.

This very earth has become a mosque and a sacred prayer carpet spread out by Allah for humanity. This very earth has become an earth of prayer, an earth of pilgrimage, an earth of fasting during the holy month of Ramadan, when the torrential rains of Divine Grace are released as a monsoon of realization. This earth has become an earth of generous giving and tender love among lovers of the Truth. This earth has become an earth of revelation, surrounded by and permeated with the Light of Prophecy. This earth resounds with the drums of *jihad,* the holy warfare of Truth, as millions of human beings awaken spiritually, affirming Unity with their very breath and heartbeat. Even the dust of the earth now purifies the warriors of Divine Love in the ablution ceremony of *tayammum,* or touching the earth, revealed by Allah through the Prophet of Completion—the truthful, faithful, merciful, peaceful, prayerful, graceful, beautiful, blissful Muhammad, upon him be peace. This Guide of mystic guides has become the Prophet for all humanity. National, cultural, and religious barriers have fallen away. There are only the lines of prayer behind the Imam of Light, aligned with the direction of Truth.

This earth is now spiritually fulfilled, according to the Promise of Allah Most High. Whose promise could be more truthful than the promise of Truth? The Distributor of the Light of Prophecy to all

Hearts without Exception, the uniquely beloved one of Allah, exaltedly proclaims a new dispensation, which has not been granted to the noble Prophets before him:

> *This earth has been made for me a place for praying and a substance to perform the purification of tayammum. Every Prophet was sent to his own nation. I have been sent to all humankind.*

SPIRITUAL DREAM

As the Glorious Quran reveals, Allah gives sleep to humanity as refreshment for body and soul. The soul's refreshment is the river of spiritual dream that pours from Paradise into the seven levels of the human heart, becoming various degrees of revelatory experience that occur during sleep, lifting sleep above physical and mental states into the realm of contemplation. All but the highest dreams require interpretation from one who is a knower of Truth. Such a person has been empowered to interpret dreams by a mystic guide, an inheritor of the spiritual wealth and beauty of the Prophet Joseph, may peace be upon him, the sublime dreamer and interpreter of dreams. The beloved Joseph's dream of stars, sun, and moon bowing before him confirmed the sublime station of the true human being, first revealed by the Decree of Allah that the angels bow before the transcendental Adam, archetype of the human soul.

An authentic spiritual dream contains one-fortieth of the power of Divine Revelation—the same Clarity and Power that descended through the Prophet of Allah as the Holy Quran. The Clear Quran discloses that were its own spiritual magnitude to descend upon a mountain instead of descending into the pristine awareness of the Prophet's soul, the mountain would be obliterated. The exaltedness of the human soul is confirmed by the fact that it can receive and bear one-fortieth of this original revelatory power, the very Power that Allah uses to create the universe with His Divine Word and to withdraw the universe with His Divine Word. One-fortieth of this Power is actually entrusted to every soul in spiritual dream. That the Caretaker of the Infinite Fields of Meaning of the Holy Quran, the beloved Muhammad, could receive the complete power of Revelation, rather than one-fortieth, confirms his adamantine soul to be the First Light, the Muhammad of Light, the Light of Prophecy, which has descended directly from the Source of Light in increasing intensity to every spiritual nation through the sublime lineage of the 124,000 Prophets of Allah.

The most exalted spiritual dreams directly perceive the Prophet Muhammad, may Allah's Peace forever surround him. Equivalent to dreaming of him is dreaming about his noble family or companions who are an intimate part of his spiritual being, or by extension, any of the Prophets, who have all borne precisely the same Light of Prophecy, or by further extension, any of the empowered shaykhs who are the inheritors of the Wealth of the Prophets. These mystic guides, who appear in every generation, lovingly sit knee-to-knee with the Prophets in spiritual experience. They have grasped the right hand of the Prophets in spiritual realization. To dream about any of these holy persons, directly or indirectly—including places and events associated with them or simply their names—not only refreshes but instructs and exalts the soul, conferring various levels of the most precious spiritual gift, companionship with Allah and the Prophets of Allah, may peace be upon them all.

The imagery of spiritual dreams can be disturbing, shocking, confusing, unclear, or even totally obscure. The inspired interpretation must fit the dream as a key fits the lock for which it was designed. The door of awareness of the dervish dreamer opens when that key is turned by the careful hand of the shaykh, the precious hand that pours the Wine of Love into the empty glass of the surrendered heart, filling it to the brim and even over the brim. The dream and its interpretation are then experienced as one reality, not as two separate events that happen to come into conjunction. Only the dream imbued with its interpretation, the response to it ordained by Allah Most Subtle, becomes one-fortieth of Divine Revelation. The dream by itself is simply a lock without a key.

The interpretation by the shaykh, or by one of the initiated representatives of the mind and heart of the shaykh, need not predict precisely future events or analyze dream symbols. The interpretation can be simply an ecstatic response, an exclamation of joy, offering praise to Allah for His Revelation, His Path and His beautiful human souls who race along this path deep into Divine Light. Or the interpretation can be a word of warning, echoing the compassionate warning of the Holy Quran to turn only toward Truth, on all levels of experience, in all areas of responsibility.

Dreams that are not spiritual, that do not emanate from Paradise, are the interior adjustments of the psyche in its attempts to process the mass of mental, physical, and emotional impressions that it contains, both consciously and unconsciously. These psychological dreams are as natural to human functioning as tears. Yet just as there are spiritual tears and sentimental tears—the difference between the two needing clear discernment by the contemplative practitioner—so there are spiritual dreams and psychological dreams. A spiritual dream usually declares itself as such to the dreamer, not by revealing its secret meaning but simply by its vibrant clarity and its total unexpectedness.

There are dreams lower on the scale of illumination than ordinary dreams of anxiety or wish-fulfillment. These are dreams inspired by demonic influence. The daily spiritual hygiene of a holy way of life and constant clarity about one's motivation—recognizing and rejecting the movements of the grasping self, the domineering ego—provide the best protection against demonic dreams, which enter the nervous system as bacteria enter the digestive system. If our daily lives are clean, clear, harmonious, humble, truthful, and merciful, these alien influences may pass through us but will find no place within us to establish themselves.

Before falling sleep, one should make prayerful preparations, for dreaming is just as much the spiritual path as waking life. Waking existence is, in fact, a form of dreaming carried on at the particular level of maturity one has attained. The same objective event has entirely different meanings for individuals at different stations along the spiritual path.

Creation as a whole is Allah's Dream, reflected through the struggle of His conscious creatures to evolve. Paradise is Allah's Dream, completely unveiled. The goal of the mystic way, the awakening into Essence, lies beyond both the dreams of this world and Paradise. May we take refuge with the Bearer of the Green Banner of Guidance, the Final Interpreter of the Dream of Manifestation, the Guide of all Souls, the beloved Muhammad. He assures spiritual dreamers who long only for Reality:

> *Whoever sees me in dream, has surely seen me, for Satan cannot impersonate me.*

MOSQUE

A mosque is not essentially a structure built by human hands but is the holy sanctuary of human existence. Divinity does not dwell in structures. The holiness of the human heart—its instinct for friendship, its hunger for the One Reality, its ecstatic response to the Call for Prayer, its thirst for the living water of the Holy Quran—is what constitutes a mosque, what builds a mosque, what loves a mosque. The prayers of the lovers of Truth, offered everywhere at the coming of the times for prayer—in mountains, deserts, green fields, villages, and vast cities—have transformed the entire planet into a mosque.

Allah Most High is not contained within any mosque, nor is Allah inside His Creation. The mosque is where this fact of Divine Transcendence is known and clearly demonstrated.

The faithful stand in congregation within the *jami,* the gathering place of true lovers, brought together by the Divine Power, *ya jami,* the Gatherer. Shoulders touching, we gaze together in the directionless Direction of Prayer, which is not a direction within the world, for we are now facing Allah alone. The mosque is perfectly transparent in the direction of transcendence. There is no city, no earth, no creature, no universe before one as one prays toward the City of Light, Mecca the Ennobled. There is no Kaaba constructed from stone in front of one who truly prays. The moment one has proclaimed the Words of Transcendence, *allahu akbar,* not even angels intervene between the gaze of the lover standing for prayer and the Divine Light of the Creator, Exalted be He. The angels may gather to the right and to the left, behind and above, but never in front. The mosque is the place that faces only Allah, the place where the secret door opens beyond all dimensions. Allah the Utterly Alone is not another dimension of reality. Allah is the Reality from which all dimensions spring, to which all dimensions prostrate, toward which all dimensions are returning, and within which no dimension can exist. The mosque is the *masjid,* the place of prostration, where all dimensions lose themselves in prostration. The mosque is the ideal human place of nonviolence, protection, healing, friendship, and illumination. Here, the forehead of the true human being touches the earth of humility in the mystery of submission, causing the brow to become as transparent as the Direction of Prayer and therefore to shine with Divine Light.

The external form of the mosque can manifest beauty and grandeur or rustic simplicity. Kings and laborers pray together. Their movements and intentions remain exactly the same. At the moment of prostration, the deepest point of the prayer, every facet of the external mosque disappears. There are no more beautiful calligraphies, graceful domes, or shining minarets. The person of prayer extends this spirit of prostration throughout the prayer. The one lost in prayer does not care if he is standing on bare earth, rough stone, smooth marble, or soft carpets. One knows then only that the mosque is the place that shines with Transcendent Light.

In this all-engulfing Light, every human consideration and opinion, all aesthetics and philosophy, disappear like stars when the full moon rises.

The Holy Mosque is the reflection of the Divine Attributes. Its ever-expanding architecture and the sublimity of its building materials do not represent human ideas, human initiative, or human self-glorification but are spontaneous expressions of Divine Power, Beauty, Light, Abundance, Harmony, and Knowledge. The calligraphy in a mosque is the Living Quran, not essentially a product of human art. The lover of Truth who enters a mosque does not meditate on any human craftsmanship there. Towering minarets and vast floating domes are not the self-congratulation of human engineering. The mosque is clearly recognized as the expression of the Will of Allah, the reflection of the Attributes of Allah. All praises are offered only to Allah.

We read in the precious Oral Tradition about the gradual expansion of the mosque in Medina, which began originally in a sheepfold. This is a mystical teaching about the progressive revelation of the Divine Attributes. Narrates Abdullah bin Umar, "In the lifetime of Allah's Messenger, the mosque was built of adobe, its roof the leaves of date palms, its pillars the trunks of date palms. Abu Bakr did not alter it. Umar expanded it on the same pattern, changing the pillars into carved wooden ones. Uthman changed it by expanding it to a great extent. He built its walls with engraved stones and made its pillars of engraved stones and its roof of teakwood." The blessed oasis of prostration, the direction of Truth, and the Divine Light, which shines from the foreheads of the lovers, remain exactly the same under the teakwood roof as in the dusty sheepfold or beneath the vast dome of the Blue Mosque in Istanbul. The quality of transparency to the Light of Allah and to the Will of Allah that constitutes a true mosque remains exactly the same and will remain so until the End of Time. Future mosques may be built of glass, crystal, or laser beams. Mosques may be constructed that are so gigantic that Mount Arafat can serve as the *minbar* and the

City of Mecca as the *mihrab*. Weightless mosques may orbit life-bearing planets.

The true mosque is always the same. What evolves is our human perception of the Attributes of Allah, Divine Energies that manifest as the universe with awesome beauty, dazzling harmony, and inconceivably sensitive mercy. This is what humanity learns to contemplate by serving as instruments for the construction of mosques. For the awakened hearts of the intimate friends of Allah, the whole universe and this precious human body have already become the ultimate mosque. But their tender appreciation flows to every small corner where the Prayers of Islam have been established with even minimal concern, care, love, and sacrifice. Divine Beauty and Mercy shine richly through these actions of prayerful human hearts and hands. Thus the Leader of the Prayers of Humanity, the Imam of the Mosque of the Entire Planet, the beloved Muhammad, reveals the mysterious correspondence between earth and Paradise:

> *Whoever builds a mosque, intending Allah's Good Pleasure, Allah will build for him a similar place in Paradise.*

TOUCH

With shoulders touching, we stand in the powerful configuration of faith, the perfectly straight lines of prayer. Just so, the radiant persons of the Resurrection will advance through the Gates of Paradise in vast lines of mutual love, each consisting of seventy thousand lovers, shoulder to shoulder. Touching and formlessly intermingling, the souls in the Universe of Souls are ranged in lines of prayer from before eternity, immersed together in the resonance of the Transcendent Quran. We are always touching, mingling, bonding, mutually supporting. We link hands as we walk toward the mosque, each step counting as prayer. We join hands as we move in the mystic circle of the dervishes in ecstatic remembrance.

The beloved one of Allah, having been sent for all humanity and remaining in subtle communion with all humanity, held a small girl on his noble shoulders as he stood for prayer. He placed her beside him on the ground as he made his prostration and then returned her to his shoulders as he stood once more to chant Quran.

We sit with our knees touching the knees of the shaykh when we take his hand to enter the mystical Order, the interlocking spiritual transmission, the touch of right hands for fourteen centuries. Divine Names flow through us at the smooth touch of each prayer bead, as we are strung together on the thread of *la ilaha illallah*. With the intimate touch of a holy kiss, we greet the sacred Black Stone, which received kisses from the lips of the Transmitter of the Living Quran, may Allah tenderly embrace him in Divine Peace. Through the palms of our upraised hands, we receive the emanations of Divine Energy flowing from the One Source of Reality through the Black Stone, as we circumambulate the Kaaba in intimate touch with our brothers and sisters, the whirling current of the faithful in union with the ones they love. With a tender touch of our lips we greet the external form of the Holy Quran, its notation with ink on paper. With this same sweet and humble kiss, we greet our shaykh, kissing the knees that have touched the earth so frequently in prayer and kissing the palms of the hands that have been so frequently upheld in intercessory supplication, receiving the direct touch of Allah's Blessing. With our own palms, energized by the Blessing of Allah during our own supplication, we touch our faces and our bodies, fusing with the Divine Touch. Our bodies touching, we sit closely together at the meetings of the dervish lovers, feeding each other with our hands, kissing each other on both cheeks, preparing food lovingly for the faithful to break their fast during Ramadan, pouring water and tea for each other, sharing water for ablutions.

With the ecstatic kiss and sensitive embrace of love, marriages are sealed and children are conceived. The mother then holds the child within her womb for nine months and feeds the child at her breast for two years, with her most intimate loving touch. What lavish human touch descends as the drenching rain of love upon children, creating for us all the experiential basis for the sacrament of touch, the loving human interdependence and dependence on Allah. All the creations of Allah, animate or inanimate, touch uninterruptedly in complete mutual interrelation—sun, grass, flocks, joyous communal meals, sheepskin posts upon which dervishes are kneeling, immersed in the affirmation of Unity, white cotton or linen of their robes, wool of their dervish cloaks, silver of their sacred medallions mined by human hands from the depths of the earth, sweet atmosphere of the planet breathed with every breath in conscious or unconscious gratitude.

We touch and reinforce each other intimately in learning and communication, the very voice of our teacher, companion, or colleague touching the sensitive eardrum. Our hearts touch and intertwine in compassionate love that actually experiences in tangible form the joy and pain of the other. The angels touch us subtly as they guide, protect, and illumine. We are touched by each other's commitments. As we advance along the mystic way of nearness to Allah, we become nearer to each other. We are more deeply touched by one another. Our sense of touch becomes universal. The universe becomes our body. We actually sense the embrace of Allah through every pore of the skin. We feel the warm sunlight of Divine Unity shining within our secret heart, just as we feel the life-giving sun streaming upon us through the cosmic creation. We will taste the fruits of Paradise together and hold the cool goblets of Paradise for each other with hands composed purely of Divine Light, their sense of touch a million times more evolved than these physical hands, which are composed of earth and Divine Light in conjunction.

Truth is a living touch among us, just as weavers can distinguish different threads merely by touch. The Divine Beauty we perceive in each other touches us with spiritual tears. The resonance of the Holy Quran touches us and unifies us into a single symphony of praise. The person who is chanting scripture, the person who is listening in rapt attention, the person who is teaching sacred tradition, the person who is learning with deep aspiration, the person who is planting seeds in the name of Allah, the person who is harvesting with the *bismillah,* the person who is generously preparing food with the *bismillah,* the person who gratefully begins the meal by invoking the name of Allah—all faithful persons whose lives are the spontaneous cry of *alhamdulillah* are intimately linked in loving, mutual touch. All streams of praise arise from the hearts and minds of living beings and flow into the ocean of the single Source of Being.

Reports the sublime Oral Tradition of the boundless Reservoir of Tenderness, Muhammad Mustafa:

> The Prophet said, *A faithful believer to a faithful believer is like the bricks of a wall enforcing each other.* While saying that, he clasped his hands together, interlacing his fingers.

PRAYER

Within the royal garden of human reality shines the white rose of the noble *sharia,* the Sacred Law, the holy way of life. To follow *sharia* is to live in harmony with creation and in communion with its sublime Creator. This way was clearly demonstrated by the lover of prayer, the lover of fragrance, and the lover of love, the beloved Muhammad, may mystic union with Divine Love always belong to him, to his beautiful family, and to those who love him. This impeccable white rose, this purity of action and intention, radiant as the gentle morning sun, grows beside the river of prayer, where the lovers plunge and swim together joyfully five times every day. Not even a mote of dust can remain upon the mirrors of their hearts.

Earthly temporality has become the five periods of prayer that follow each other gracefully, as the green earth rotates and revolves within the harmony of the Cosmic Quran, whose *ayats,* whose signs or verses, constitute the creation. Hours of sun time are now revealed as hours of prayer time, and the universe is disclosed as an instrument for the supreme music of prayer. The white light that spreads across the horizon is the delicate presence of Dawn Prayer, the perpetual dawn of aspiration. The zenith of the sun's arc is the powerful presence of Noon Prayer, the natural uprightness of the human soul. The slanting afternoon light is the sustaining radiance of Afternoon Prayer, the Divine Grace of continuous spiritual nourishment. The twilight that blossoms fragrantly after sunset is the tranquility of Evening Prayer. The rich darkness, radiantly black or shining with moonlight or starlight, is the mystery of Night Prayer, which prepares holy lovers for the midnight sunrise of mystic union.

The Man of Perfect Prayer, upon him be peace, prayed many more than the five basic prayers of Islam. Affirms his youngest wife, the jeweled light of his eyes and the sublime transmitter of one third of his Oral Tradition, Aisha, Mother of the Faithful, may Allah ennoble her in mystic love: "The Prophet used to remember Allah at all times." Nevertheless, within these five brief periods of prayer, the entire range of spiritual experience is expressed perfectly each day. Similarly, within the graceful movements of a single cycle of prostration, or *raqat,* all forms of angelic praise occurring throughout the heavenly realms are joined together harmoniously and become the vehicle for the mystical ascension of the faithful.

Dawn Prayer is the sweet submission of the soul to Allah alone. Noon Prayer is the responsible and beautiful life of courageous action that offers all praises only to Allah. Afternoon Prayer is the yearning of the soul and its struggle to persevere on the path. Evening Prayer is the resurrection into the Garden of Paradise. Night Prayer is the intimate nearness to Allah that prepares the soul for the mystic union of lover and Beloved, the culmination of the path, the blissful return of the ray of light into the Source of

Light. These are drops from the limitless ocean of the spiritual meanings of the Five Prayers, the shining white banner of the noble *sharia* of Islam.

The melodious Call to Prayer pervades all creation, flowing through the instrument of the human voice from the essential silence of the supreme Source. This invitation to conscious unity awakens the ecstatic response of prayer within the hearts who tremble inwardly at these resonant words of the Call to Love, the Call to Realization. Leaving the entire complex world of the limited self behind without the slightest regret, the invited guests stream to the banquet chamber of the King of Light. They gather within the Palace of Vision so vast that no one can perceive its walls. Intoning the Words of Transcendence, *allahu akbar,* which bring our awareness beyond limited concepts and partial experience, we face the Direction of Prayer. This is the straight path of Light, stretching directly to each heart from the full moon of Truth, reflected across the waters of creation.

As we raise our hands above our shoulders, by the Power and Permission of Allah, relative existence falls behind us. With the seven mystical steps of Sura Fatiha, containing in condensed form all revelation that has ever come to humanity, we begin to ascend along the path of Light to the formless goal of Light. Bowing at the waist, we gaze deep into the inward glory of Allah at the heart of our being. Plunging into prostration, we lose ourselves in the Ocean of Love. Seated in the kneeling position, we abide in joyous companionship with Allah the Beloved and the 124,000 holy Messengers of the Beloved—from Abraham, upon him be peace, to the Final Fruition of the Light of Prophecy, the Rose Garden of the Nightingales of Quranic Verses. Finally, imbued entirely with the Peace of Allah, having become overflowing goblets of prayer, we offer Divine Peace to our right and left along the prayer line of all human beings who have ever prayed or will ever pray to the One Reality, Who receives all prayers and Who continuously receives the entire universe back into His Essence, exalted beyond all conception be He. This beautiful offering of peace at the culmination of prayer is received gratefully by all the kingdoms of life as a sweet spiritual fragrance.

Opening our hands in pure supplication, not grasping at any finite desire in this world or the next, we receive through our palms the particular energies of blessing that Allah Most Subtle ordains for each person at each time of prayer. Placing our sanctified palms upon our faces—as the Warrior of Truth, the noble Moses, placed his right hand beneath his arm—we shine with radiance for anyone who has Allah's Permission to perceive the *nur al-islam.* Light of prayer now beams from the foreheads of the People of Prostration, the People of the Right Hand, the People of Paradise, the People of Peace, the People of Nearness.

These are atoms from the boundless sun of the radiant meanings of the movements, positions, and spiritual stations of the Prayer of Islam, brought down by the holy Prophet from his magnificent Night Journey through Divine Attributes into Divine Essence. With what ecstatic longing, wonder, and awe should we approach the lines of prayer, which are twenty-seven times more powerful than the same prayer conducted in solitude. May we always move with the graceful movements of the prayer and the silent dignity of the prayer, for prayer is the very nature of our humanity. The beauty of our person, including the beauty of our physical form, is only the beauty of prayer. We are always waiting for the next time of prayer, and that waiting is counted as prayer. We are prayer, coming to prayer and abiding as prayer. We are prayer serving, prayer sleeping, prayer breathing. Thus the Leader of the Prayers of Humanity, the One Who is Perfected in Prayer, instructs those who long to pray to come to prayer with the human dignity and Divine Peace characteristic of Islam.

The precious and illuminating stream of the Oral Tradition recounts:

> While we were praying with the Messenger of Allah, he heard the noise of some people approaching. After the prayer, he asked them

why they were making such commotion and they replied that they were hurrying to join the lines of prayer. He said, *Do not make haste or run to prayer. Whenever you come for the prayer, you should come with calmness and solemnity.*

BISMILLAH

Through beginning each action, taking each step and breath, following each train of thought, and formulating each intention with *bismillah ir-rahman ir-rahim*—in the Name of Allah, All-Compassionate and All-Merciful—the dervish lovers who gaze into the mirror of spiritual dream will see that their heads have disappeared into graceful flames.

The musical cry *bismillah ir-rahman ir-rahim* will then resound from the entire being of the dervishes and fill the universe. From the shining eyes of such persons, tears of love will fall like huge raindrops upon the white marble floor of the Grand Mosque in the Medina of the heart. The Mercy and Compassion of Allah will flow through the open channel of such persons with great subtlety, gentleness, and profusion. Each thought, perception, and action occurs no longer simply in the Name of Allah's Mercy but as the very Mercy of Allah. This is why the holy and tender Prophet, upon him be peace, is honored as the Mercy of Allah to all the Worlds. He actually became *bismillah*. He breathed only *bismillah*. He slept and ate, walked and talked only through and as *bismillah ir-rahman ir-rahim*.

This spiritual station of becoming *bismillah*, this full realization of the Compassion and Mercy of Allah in and through one's entire being, was transmitted by the Prophet to the holy ones of his community. Merciful lovers of humanity from his pure lineage of Divine Compassion and Mercy now exist everywhere across this green earth, like sacred, ever-flowing springs, living lives of mercy alone.

Allah is Mercy, the Prophet of Allah is Mercy, the religion of Allah is Mercy, the devout practitioners of the religion are Mercy, the followers of the mystic way are Mercy, and the goal of the mystic way is to merge entirely into infinite Mercy. The human bearer of Divine Mercy never so much as tears a green leaf from a tree, as the merciful Muhammad teaches in his sublime Oral Tradition. The thirst of a dog eating damp sand around a desert well becomes our own burning thirst, and we fill our shoe with water so the suffering animal can drink, as one of the intimate companions of the Prophet is recorded to have done.

Through the beloved Muhammad—Intercessor for Humankind, Jinn, and Angels and Opener of the Floodgates of Mercy into the Field of Hearts—the human bearer of Divine Mercy receives the spiritual gift and continuous function of intercession. Those who have received the gift of intercession through the holy Prophet and his spiritual lineage, passed from hand to hand along the line of mystic shaykhs, pray only for others. Such persons simply become mercy. Their only longing is that human beings fall in love with Divine Unity and enter the Paradise of Unity. These persons of mercy never consider entering Paradise until all beings have been removed from the anguished thirst of negativity and have drunk their fill from the clear fountain of the affirmation of Unity. If the thirst of a dog spurs the true lover into compassionate action, how much more will the spiritual thirst of human beings, who are all our beloved ones, elevate us to the highest level of action—a life of constant intercession in the intimate abode of the All-Merciful One. This way of life will continuously proclaim Divine Mercy, living and breathing *bismillah ir-rahman ir-rahim.*

The Good Pleasure of Allah, the Consent of Allah, the Sweet Response of Allah, and the Mystic Embrace of Allah are directed to the person of mercy. A single tear of pure love that falls from the corner of the eye of the person of mercy puts out the fires of hell. An instantly outstretched hand of selfless compassion is more pleasing to Allah than all the prayers of a lifetime performed with self-involved and self-serving motivation. A single drop of loving-kindness, falling as miraculous rain from the Paradise of self-forgetfulness and continuous remembrance of Allah, can purify, sweeten, and make fragrant an entire lifetime, an entire family, an entire community—as a potent drop of rose essence can make a hundred vials of perfume, bringing refreshment and delight to many persons covered with the dust and cares of the world. The sweetest Mercy of Allah that flows through the person of mercy is often hidden, or tiny as an atom, expressing Divine Mercy perfectly through small acts of kindness, which mysteriously contain the boundless Power of Allah, the Power that creates and withdraws the universe. This energy of Divine Mercy, vibrant within our hands and within our intellect, prepares and clears the way for others, as Allah Most Merciful prepares the earth for life, opening fertile soil and wombs with seeds of spiraling growth. Through this energy of Divine Mercy, plants, animals, and human beings awaken together in harmony.

The person of mercy, the embodiment of *bismillah,* exercises utmost intensity of attention at every moment: visiting prisoners, nursing the sick, freeing those who are enslaved, feeding the hungry, educating those who hunger for knowledge, defending the oppressed, consoling the bereaved, protecting orphans, resurrecting hearts that have spiritually died, singing out with hearts that are alive in the Living One, demonstrating the life of praise to new generations, rekindling courage and faith in the hearts of the grandmothers and grandfathers. As the single sun transmits abundant life to a vast variety of living beings, by the Permission of Allah the person of mercy shines forth from Divine Mercy as rays of tender compassion during every second of his or her existence—a truly human existence, all its dimensions drenched with *bismillah,* overflowing with *bismillah,* fused with *bismillah.*

How sweet a single word or glance from the person of mercy, within whom and from whom the radiance of Allah's Own Delight is streaming! The person of mercy pours water for our ablutions, leads us in selfless prayer, and brews us delicious coffee. The person of mercy, acting solely in the name of Allah Most Merciful and Compassionate, wipes away our tears of sorrow, heals our diseases, arranges for our dwelling and our livelihood, removes obstacles from our spiritual path, invites us again and again to Truth, and brings us to tears of laughter and ecstasy. The person of mercy acts invisibly, preparing our way before we come and cleaning up after we leave. Divine Mercy is the invisible atmosphere of such persons, which we breathe in their presence without being aware. The constant compassionate acts of these merciful ones are swifter than eye or thought. These beloved friends of Divine Mercy are the Ocean of Mercy, and we are fish, freely swimming there.

Allah Most Merciful calls every human being to be merciful in precisely this miraculous way. This call to mercy is Islam. Every human soul contains infinite treasures of Divine Mercy, locked away behind door after door of the limited self, with the locks of limited concepts and limited aims. When we call out *bismillah ir-rahman ir-rahim,* all these locks are shattered and these doors swing open, falling away forever from their hinges. At this Cry of Mercy, *bismillah ir-rahman ir-rahim,* the release of life-giving love is greater than the ancient Flood with which Allah once cleansed the entire earth. This secret, melodious cry of *bismillah,* deep within the heart, is the very Resonance of Allah, the mystic trumpet sound that will stop the universe at the End of Time. The *bismillah* brings our narrow world of negation and imprisonment to a sudden standstill, removes all barriers, opens all doors, floods the precious human heart with the subtle light of Divine Mercy, and creates Paradise on earth.

The Fountain of Compassion, the Blossoming of Full Humanity, the Resurrection of Love, the merciful Muhammad, may the Mercy of Allah surround him always, compassionately reveals in his Oral Tradition:

> *While a man was going on a way, he saw a thorny branch and removed it from the way. Allah became pleased by his action and forgave everything for that.*

SUBMISSION

Whatever event that may spring from the Source of Being has been written from eternity in the Transcendent Quran. Even the demonic energy of negation and rebellion operates throughout creation and whispers in the outer courtyard of the human heart only by the Permission of Allah, as testing, strengthening, and teaching for the soul. Those who witness to the Truth of Islam with the force, conscience, and integrity of their entire being affirm that whatever happens, whether it appears helpful or harmful to the eyes of our limited understanding, is sent justly by the Source of Wisdom to refine our faith and to reveal our level of submission to the Divine Will. The great lovers of Truth, the mystic shaykhs, are like lightning rods continuously being struck by the Lightning of Allah. They take calamities upon themselves, by the mysterious Permission of Allah, that might otherwise have fallen upon their students or upon humanity as a whole. The realized lovers of Allah who are not suffering disasters become concerned that their Lord no longer honors them so deeply by trusting them to bear burdens for others.

The dervishes who walk the mystic way not only accept but cheerfully welcome the sufferings that come upon them by the Will of Allah, regarding every form of privation, pain, and struggle just as the sword in the hand of the weapon maker regards anvil, hammer, and fire. The more intense the suffering, the more precious the opportunity for refinement of faith, increase of wisdom, purification of heart, and blossoming of selfless patience, that Gift from Allah so sweet that the human soul who tastes such patience needs nothing else. For human reality to be transmuted into a clear mirror for the Divine Attributes—Patience, Faith, Justice, Beauty, Wisdom—more pressure is necessary than required to transform coal into diamond. To welcome the overwhelming pressure that Allah Most Awesome brings upon the soul through internal experience and external events, never burdening the soul with even one atom's weight more spiritual responsibility than it can bear—to welcome this pressure instantly and instinctively, with all one's heart and mind, is submission.

Submission is the advanced training of the soul. One submits to Allah often with tears but always with spiritual joy. Any obvious or subtle presence of emotional heaviness, complaint of spirit, or clever pattern of avoidance and self-justification invalidates our submission and brings us back to the beginner's level of accepting the Will of Allah reluctantly and under protest. Authentic submission to Allah Most High creates the sensation of effortlessness. It is sweet and graceful, accompanied by the deep inner silence that indicates the increasing transparency of the layers of the limited self to the Light of the limitless Source. The great efforts made to submit personal willfulness to the All-Embracing Will can never succeed. No finite religious acts or moral disciplines can guarantee a place in the infinite Paradise of willessness. Pure love for Allah alone—possible only because Allah mysteriously reveals His Love for us and places this Divine Love within us—can bring the sincere lover to the gates of the Garden of Submission. No one can open these gates other than Allah the All-Opening.

Submitted persons undertake no thought or action on their own. The Glorious Quran reveals how the illumined Friend of Allah, the hidden sage Khidr, conducted the noble Moses through the elevated regions of submission, disclosing that even the Messengers of Allah must study in the highest rank of the university of souls called submission. Beyond all levels of the mystical path, even beyond the attainments of masters who have completed the path, lies the highest station, the *islamiyya,* the perfection of submission. Submission is absolute transparency to the Will of Allah. There is no longer any separate subject who is submitting to any Divine Force, imagined as imposed from above. There is now only the Will of Allah. There is no compulsion, no one to be compelled. The soul has not disappeared into *fana,* ecstatically merging into Allah, but has been established in the indestructible diamond state of *baqa,* moving in mystic companionship with Allah yet remaining entirely within Allah as nothing but an open channel for the expression of Allah.

The person of submission becomes a conscious embodiment of Divine Will. Outwardly, this person appears exactly like other human beings—decisive, exercising strong preferences, holding high standards. Inwardly, this submitted one is vast emptiness, filled dynamically, moment by moment, with Divine Radiance and Power. The serenity of this holy emptiness surpasses all understanding. At first the dervish strives to be an empty vessel. Now the striving and even the vessel have disappeared into Allah and reemerged as Allah. The clarity and merciful energy of this boundless openness of submission is Allah's Own Delight and Allah's Own Creativity. Subtle beauty and startling, unexpected simplicity shine from the person of submission to everyone whom Allah gives permission to see. Divine Light plays about this person in the midst of ordinary situations. The person of submission could be mistaken for an old bookseller or an ice-cream maker.

The one who is pure submission, the beloved Muhammad, will shine the delicate light of submission upon the whole world until the End of Time. He is the full moon of tenderness that never wanes and the sun of wisdom that never sets. The sublime Oral Tradition reports that when asked about the daily life of the holy Prophet, the precious Aisha, jeweled light of his eyes and Mother of the Faithful, responded in words that scintillate with the light of submission: "He used to keep himself busy serving his family, and when it was the time for prayer, he would go for the prayer." Gaze at the mysterious simplicity and dignity of submission, dear brothers and sisters, shining through another *hadith:*

> A cloud came and it rained till the roof started leaking. In those days, the roof of the mosque used to be the branches of date palms. The invitation to prayer was chanted, and I saw the Messenger of Allah prostrating in water and mud. I even saw the mark of mud upon his forehead.

LIGHT

Opaque curtains, translucent veils, clear windows, open doors, bright mirrors, brilliant lamps, windowless and doorless barriers that open mysteriously, bridges of radiance over precipices and oceans, broad highways of sunlight, narrow paths of moonlight, beams of radiance emanating from the breast of holy beings on moonless nights, lightning flashes prolonged in ecstasy, angels of Light, ladders of Light, and stairways of Light—these are the landscape of the mystic way. Spiritual travelers encounter these forms of Light, which take form temporarily and change form instantly, as they traverse the infinite dimensions of the unified field of Light in bodies of Light drawn to the single goal of Light.

All existence, as the hidden depth of the Glorious Quran reveals most mysteriously, is the Light of Allah within the Light of Allah, the *nurun ala nur.* The cry of the soul is therefore "Light, light, more light, and only light." The Light of Allah is the illumination of heaven and earth, the only medium of awareness and being, the sustenance of souls. All creation longs for Light and evolves toward Light. We are fish swimming in the ocean of Light, birds flying in the sky of Light. The insubstantial shadows of negation are cast only by the single Light of Affirmation. Atomic structures of matter, molecular structures of life, linguistic structures of culture, and ecstatic structures of spiritual vision are complexities that consist only of Light and subsist only within the essential simplicity of Divine Light.

All beings in the vast theater of Divine Creation are waiting for the curtain to rise, revealing the ultimate stage of Light. The whole universe is poised, waiting for the call of Light. The lovers of Truth are deep in prayer, waiting for the smile of Light. The beautiful letters of the Arabic Quran are written by the pen of Light upon pages of Light with the luminous ink of human awareness and memory. The awesome letters of the Cosmic Quran are the billions of stars, bearing the fruit of life-sustaining planets and flowering civilizations that contain precious beings of Light who turn in prayer only to the single Source of Light. There are eighteen thousand universes, each an ocean of countless galaxies—swirling currents of Light containing whirling dervishes of Light.

The First Light, *nur muhammad,* the Light of Prophecy that streams through all civilizations, stands in the open door of Light in his abode of Light, gazing with his gaze of Light upon the lovers of Light, granting them the supreme gift of his smile of Light. With his strong right hand offered to those who have stumbled or fallen along the path of return into Light, he holds up the door curtain of Light, revealing himself for a few moments of brilliant Light in his humble dwelling in the ancient desert wilderness of Arabia. Then he lets the curtain fall again, and the Light of his countenance, so tender and beautiful, no longer shines externally but springs forth as a fountain of Light in the luminous Paradise of the hearts of his lovers.

Allah Most High, exalted beyond all conception be He, created the eighteen thousand universes solely for the Muhammad of Light and through the prismatic instrument of the Muhammad of Light—for *nur muhammad,* through *nur muhammad,* to *nur muhammad.* The entire spectrum of energy, space, time, and eternity is Light alone, appearing to be separated into kaleidoscopic wavelengths. There is no actual separation within Light, nor is there any other reality for the One Light to be separate from. The great wave of revelatory Light upon this earth, which began with the Prophet Adam, may the radiant Peace of Allah shine upon his soul, has reached the shore of Reality and has culminated as the sparkling white water of Light, the bright wave falling in prostration at the Throne of Light, the beloved Muhammad of Arabia.

Now all creation must become, through the Power of Allah alone, one conscious wave of Light containing all lives, all energy forms, all souls. At this moment of culmination, the Muhammad of

Light will drop the curtain upon the luminous pulse of creation, as he dropped the curtain for the last time in front of his shining human face. The explosion of inward Light will know no bounds. Curtain, wall, doorway, desert, and prayerful lovers will disappear into Light and reappear as Light, where nothing is lost and where all souls are with the ones they love.

O translucent Pearl of the Universe, O Muhammad of Light, by the mysterious Permission of Allah, the curtain is in your hand. Through you, the original Treasure of Essence has been revealed as Light. For you, the Source of Light created the creation as vast structures of Light. Through your Light, these structures have become manifest, and by your Light of Prophecy they have reached their culmination. Your Light, *nur muhammad,* is now calling all beings of Light to return consciously and freely into the Source of Light. Your Light, from which souls and universes began, will now end the universe as well, embracing all into mystical union. Gracefully, you are standing in the doorway of Light, gazing for a moment with infinite tenderness upon humanity, smiling and holding the door curtain.

As we read in the radiant Oral Tradition:

> When the people were aligned in rows for the prayer, the Prophet lifted the curtain of his house and started looking at us. His face gleaming like a page of the Holy Quran, he smiled with joy. Abu Bakr retreated to join the first row, since he thought the Prophet would come to lead the prayer. The Prophet beckoned us to complete the prayer, and then he let the curtain fall. On that same day, he died.

MOTHERS

Mother of the Living, the beloved Eve, may the Peace of Allah always be upon her, received on her gentle brow the shining Light of Prophecy after the passing away of the first holy Prophet on earth, the beloved Adam, may the Peace of Allah always be upon him. The All-Forgiving One turned His Glory toward our original father and mother, washing their hearts in the purifying and healing stream of *la ilaha illallah,* fulfilling their longing to experience mystic union, and focusing His Divine Light and His Divine Word through them for the guidance of all future humanity.

From the sacred brow of our holy mother Eve, the Light of Revelation descended through countless generations of Prophets until the appearance of the second supreme mother of humanity, confirmed in the Glorious Quran as a channel of Divine Love to all human beings until the End of Time, our sublime spiritual mother the beloved Mary, may the Peace of Allah always be upon her. Mary the Illumined courageously accepted the virgin birth of the *ruhullah,* the very Spirit of Allah, the beloved Jesus, may Divine Peace embrace him and be transmitted through him to all his lovers throughout history.

From holy mother Eve to holy mother Mary, who both received the full Light of Prophecy, the Divine Drama of revelation and mystic return unfolded progressively and was brought to the threshold of completion—a process enduring for untold thousands of years and intimately involving millions of tender, strong, loving, patient, insightful, and loyal mothers. The final threshold was crossed, five hundred years after the Virgin, when our most delicate and sublime mother, Amina the Blessed, may Allah always ennoble her countenance, received the white dove of the compassionately descending soul of the Seal of Prophecy. All beings, including the very planet, trembled in ecstasy, instinctively and intuitively recognizing the culmination of revelation, the closure of the circle from the preeternal Muhammad of Light to the Bearer of the Wisdom of all Prophets, the beloved Muhammad of Arabia.

Then great mothers and guiding lights of the faithful began to appear, like full moons sailing to the earth and living here with unimaginable feminine grace and power. This manifestation began with the first mother of the community of Islam, the first initiate to taste the intense sweetness of submission to Allah transmitted through the Prophet of Allah, the first one of her people wise and courageous enough to witness to the prophethood of the trustworthy one whom she loved and embraced as her husband, the precious Khadija, may Allah Most Glorious always confirm the queenly nobility of her soul.

Among the incomparable wives of the Prophet, each of whom is a mystic gem adorning the Green Turban of Complete Realization, the most precious Aisha shines forth as the crest jewel of spirituality, the one who demonstrates the perfection of womanhood to all humanity, may Allah eternally confirm her station at the right hand of the Messenger. This Mother and Guide of the Faithful transmitted the intimate and inconceivably powerful content of one third of the Prophet's sublime Oral Tradition. As the beloved Muhammad, immersed in prayer and lost in the bliss of mystic union, would prostrate on his bed toward the Kaaba during the secret midnight hours, our holy mother Aisha would move her legs to make room for his radiant head and then stretch out her legs again whenever her tender husband rose from his prostration to continue the sacred recitation of the Resonance of Allah, the Book of Reality clothed in the letters of the Arabic alphabet.

Finally, the manifestation of Divine Attributes through the feminine form of humanity reached its culmination in the one who does not fit into any books or words, the majestic Fatima, may the secret of her union with Allah and with the Prophet of Allah be revealed and replicated in all mature hearts. The noble Messenger proclaimed unequivocally of the august Fatima, "She is part of my prophecy." The sublime Ali, Whirling Lion of Allah, most mature among the spiritually mature, recognized our holy mother Fatima as the inward successor of the Prophet of Allah in the mystical lineage hidden within the secret heart of Islam. He therefore did not take hand with the first Khalifa, Abu Bakr the Truthful, until, after six months of unimaginable yearning, the brilliant light of the soul of Fatima left this surface world to join the soul of her father in the Garden of Essence, where there is only one soul.

Our most profound and humble greetings, salams, and kisses to the earth where the feet of these holy mothers have walked: beloved Eve, beloved Mary, most precious Khadija, most precious Aisha, and the Pearl Beyond Price—the majestic and mysterious mother of the Mirror of the Prophet, the noble Hussain, may the Peace of Allah always be upon him, the Mother of the Ecstatic Lovers, Supreme Lover and Beloved of the Prophet—Fatima the Enlightened.

The mothers of all humanity are reflected in these radiant Mothers of the Faithful. When asked who is the most important person for the soul, the noble Prophet responded thrice, with decisive intensity, by repeating the sacred word *mother*. The Holder of Spiritual Secrets reveals in his Oral Tradition the mysterious words, "Paradise abides at the feet of the mothers." The holy tomb of Fatima the True Secret, contained in the house of the Prophet that is embraced within the Grand Mosque in Medina the Illumined, surges with a flood of spiritual power, which inundates the earth with subtle blessings. This radiant energy of love, flowing from Fatima's fragrant resting-place, is not separate in any way from the *baraka,* the transforming holiness streaming from the Tomb of tombs, the refreshing Palace of Love's Resurrection, the resting-place of the beloved Muhammad. This oasis of Love is also the destined resting-place of the beloved Jesus, after he returns and reigns over the entire globe. Here the two Prophets of Love and intimate spiritual brothers in Divine Love will manifest side by side on earth, as they do now in the highest Circle of Love.

O mothers everywhere, throughout time and eternity! Bearers of Love! Full moons of Divine Love, reflected through the precious feminine form! Most important persons to the soul! Keys and doors to Paradise! Foundation of Paradise! Please listen, beloved sisters and wives, to the sweet music of the Oral Tradition. The Messenger of Allah here discloses the tenderness of his own mother-heart:

When I stand for prayer, I intend to prolong it, but on hearing cries of children, I cut it short, as I dislike to trouble the mothers.

May the eyes of the heart of both men and women be opened to the inspiring spiritual reality of motherhood, the startling spiritual implications of motherhood, the enlightening spiritual secrets of motherhood.

The mystery of motherhood shines as Mecca, Mother of Holy Places, radiates as the Glorious Quran, Mother of Scriptures, and illumines the entire universe as *umma,* the spiritual community, Mother of the Lovers of Truth. The mystery of motherhood sparkles secretly as the primary Divine Names, *rahman* and *rahim,* which derive from the single Arabic root meaning *womb.* The mystery of motherhood glows delicately as the spiritual pregnancy of the heart of both men and women along the mystic way. This rich spiritual mystery manifests in a special sense through all women.

O Matrix of Existence! O Birthgiver, Educator,
and Protector of all Worlds! O Glorious One!
amin amin alhamdulillahi rabbi-l-alamin

LOVE

Dervish lovers on the Way of Love experience Paradise through the transparent medium of earthly life. This is the highest teaching and our essential longing—love, love, and only love. The abundant fruits of Paradise are simply love. The rivers of Paradise—clear water of awareness, pure milk of knowledge, strong wine of ecstasy, and clarified honey of mystic union—are rivers only of love. The companions of Paradise are companions of love. The perpetual springtime shade of Paradise is the incomparable refreshment of love. The goblets of Paradise are composed of the crystal of peaceful love and the silver of brilliantly flashing love in harmonious conjunction. The wisdom maidens of Paradise are luminous pearls of perfect love, scattered in unimaginable profusion. Contemplating them, the soul is elevated to the realm of Allah's Own Delight. Conversation and communion among the People of Love travels directly from heart of love to heart of love. This profoundly clarifies the mystery of our earthly experience, which is a testing and teaching for the soul, and resolves with the power of Divine Love every twist of love or hesitation to love in the entire being of the lover.

Paradise is the resurrection of the mundane body and mind as a radiant person, now composed entirely of Divine Love. Thus the dry wood of the limited, conventional, selfish body is transformed by the fire of love into the eternally flaming flower of the spiritual body. The healing and awakening Resonance of the Holy Quran is the very sound of Divine Love—the wind of the Holy Spirit blowing through the fragrant trees of Paradise, swinging their fruit-laden boughs in the graceful, dignified movements of the dervish circle of love, revolving in love to the sound HU HU HU, Essence of the Essence, or HAYY HAYY HAYY, Divine Life overflowing as all lives in creation. The banquet of Paradise is the banquet of love at lavishly spread tables of love that descend directly from the Source of Love. We are surrounded by the lovers of Divine Love, feeding each other the food of love, with the tender intimacy of those who have disappeared completely into love. We gaze at each other intently with eyes of love that perceive only love.

The circle of the highest companions of Divine Love floats as far above Paradise as a constellation above the earth. One of the central stars in this constellation of love is the Sultan of Love, Pir Muhammad Nureddin Jerrahi, may the disappearance into love and the reappearance in a body composed entirely of Divine Love—the supreme state of love, which he demonstrated on earth—be his throughout eternity. The pivotal star in this pulsing constellation of mystic love is the one who is all love, the Lover of Humanity, the uniquely beloved one of Allah, the Muhammad of Love. Next to this incomparable prophetic soul of love shine the most brilliant stars—the noble Jesus and his holy mother Mary, whose child was conceived miraculously through the virgin birth of pure Divine Love. There among the highest companions shines the victorious Muzaffer Ashqi, one who belongs only to love and is always intensely in love, along with other mystic poets and singers of love, illumined by light from the radiant star, the sweet singer of Israel, beloved Prophet David, may peace be upon him as he perpetually sings psalms of love before the Throne of Love.

The Prophet of Love, Muhammad the Beloved, promises humanity, by the Permission and Power of Allah, the most sublime among promises: "You will be with the ones you love." So the dervish lovers of Divine Love, drunk with the Wine of Love, cry out constantly with love for these highest companions of love, longing only to shine with unspeakable love in that glorious parliament of love where the Prophet of Allah, with his last precious earthly breath, prayed to be taken.

As the soul approaches that Core of Love, the forms of Paradise become increasingly transparent. The essential formlessness of Divine Love is revealed. The clear glass of the soul's lamp disappears from view as the flame of love burns brighter and brighter. The Paradise expressed through earthly imagery implodes and disappears as the love of the lovers approaches the infinite intensity of Divine Love. Then all thresholds of manifest Being are crossed over simultaneously, and the Garden of Essence shines alone as the Essence of Love.

To the purest lovers, this earth is known as the Garden of the Perfect Creation of Allah. Paradise is the secret Garden of Realization, hidden behind the Garden of Creation and discovered here and now by souls traversing the maze of swirling atoms and thoughts, holding onto the hems of the robes of the masters of love. The Garden of Essence is the Secret of secrets, hidden behind the Garden of Paradise. From this essential bower, this beautifully blossoming cloister of Pure Love, the perfected soul can look out upon the Garden of Paradise and the Garden of Creation, moving easily between earth and Paradise, as the fish of love swims from one bank to the other in the river of love. There is no division or separation.

O sweet freedom of swimming in the essential current! On one holy bank, the earthly lovers of Truth are meeting in the dervish tryst of love, and on the other holy bank, the heavenly lovers, released entirely from the limits of earthly life, are exploring and contemplating the boundless realms of Paradise. But in the middle of the streaming Essence, at the hinge of the swinging door between time and eternity, at the center of the spinning wheel of earthly and heavenly manifestation, there is love alone—without form, without subject or object—bathing both worlds with its thirst-quenching, life-giving freshness.

The Here and the Hereafter are love. Their single taste is love. As the enlightened Prophet David sings, whether we plunge to the depth of the ocean or rise to the high point of all creation, we will encounter only Divine Love. The angels moving instantaneously from realm to subtle realm on wings of light are love, and the birds with instinctive praise opening their wings for flight are love. There are no boundaries between the many currents of love within the single ocean of love. Within the Essence of Love, however, there are no shores, even, no currents, no ocean, no surface, and no depth. The two worlds, temporal and eternal, are beautiful silk scarves of love being offered openly in the bazaar of love. For a few gold coins of love for Allah, these scarves can adorn the beauty of the precious human form. But the Essence of Love is an immeasurable diamond, offered secretly by the few mystic shaykhs in the bazaar, hidden among booksellers and ice-cream makers, unknown to the jewelers, who are blinded by pride of possession, by the idolatry of collecting the precious stones of multiplicity. No amount of gold can confer ownership of this diamond of Essence, for there are no owners, there are no buyers or sellers, there are no souls separate from It. Essence cannot adorn the human form as silken scarves can, for It is formless, and no form can exist beside It. Yet from this facetless diamond alone shine forth the faceted realms that we call creation and Paradise.

To this Essence, the holy Prophet secretly alludes in his sublime Oral Tradition. After entering the Garden of Essence, the human bearer of the Essence of Love, the beloved Muhammad, may he simply be Divine Love, came forth from prayer in a body of love and proclaimed to his companions of love:

I was shown Paradise, and I wanted to take a bunch of fruit. Had I brought this fruit, you would have eaten from it as long as the world remains.

SURA FATIHA

Each noble Messenger is created by the Mercy of Allah before eternity as a sublime prophetic soul, and each comes forth, called by the Word of Allah, at a precise moment in the Divine Drama of progressive revelation. The Prophets are like harmonious notes in a vast symphony. Yet these mystic tones have resounded and will resound throughout eternity, unlike linear notes that are struck, sound briefly, and disappear again. This revelatory symphony of prophetic voices is the very Word of Allah, which called forth the universe during six original Days of Power and which is now manifest as the music of the spheres, the melodious hum of the entire creation in constant praise of its Creator, singing through the languages of all beings. This Word of Allah, which called creation instantly into being with the Divine Cry *Be!* will become the soaring trumpet blast that throws the entire universe into the stillness of trance on the Final Day. The supreme crescendo of the Word will then awaken the entranced universe into the Resurrection of Love, to live Divine Life alone, with no created limits.

What is this symphony with 124,000 movements, each marked by the unique theme of a prophetic soul? What is this Resonance of Allah, from which all worlds are coming forth and into which they are returning again, like bubbles in a stream? What is this Divine Word that creates, sustains, and dissolves? What is this Word that shines as healing and illuminating wisdom from the hearts of the Prophets of Allah and those who love the Prophets? This Resonance, this Word, this universal music of Power and Mercy is the Transcendent Quran.

The Transcendent Quran is like the sun, millions of miles above the earth and yet intimately sustaining the entire ecology of the planet with its life-giving radiance. To meditate upon the notation of the Arabic Quran with ink on paper is like gazing at the solar disc in a clear sky. It is too bright to look at directly, but it can be seen as a whole. The Transcendent Quran graciously appears as a finite book of several thousand verses, just as the unfathomable energetic depth of the sun appears as a small brilliant golden disc, floating in the blue sky. Were we to come close to the radiant sun of the Transcendent Quran, our modes of perception would be blinded and silenced by its dimensions and by its power. Were we to come closer than that, we simply would not survive as individual, personal beings. However, as Allah Most High keeps the earth just close enough to the sun to create the most fruitful environment for diversity of life, so Allah Most Subtle keeps humanity at the right distance from the Transcendent Quran, simultaneously veiling and revealing—emanating Divine Light, which does not burn the tender shoots of faith yet sustains vast ancient forests of mystical knowledge.

Each holy Prophet, may the Peace of Allah be upon them all, is given a special sign from the Source of Power and Love to demonstrate prophethood to his people. The Confirmation of the Signs of all Prophets, the beloved Muhammad, brought forth in harmony the spiritual secrets of all previous Messengers as the infinitely condensed form of the Arabic Quran, which is the reservoir of every drop of the torrential rain of revelation that has poured down upon the universe since the awakening to prophethood of the beloved Adam.

All sacred books of humanity are contained, confirmed, and safeguarded in the embrace of the Arabic Quran. The immeasurable Sun of Divine Power, the Transcendent Quran, miraculously appears as a golden disc, as an illuminated Arabic manuscript that the lovers can kiss and rest their heads upon, shedding tears of love. The Arabic Quran is the unequivocal sign brought from Allah by the Prophet of Allah. No one can imagine the power of this sign, which will shine like the sun in the sky of human awareness until the End of Time, for it is the culmination of the Drama of Revelation. Were the Transcendent Quran to descend upon a great mountain, it would be like the sun descending to touch the green earth: the mountain would be blown away like a mote of dust; the entire planet would be vaporized. This is precisely what will happen on the Final Day, when the Transcendent Quran descends to earth.

However, the flaming sun of the Quran, the full Light of Revelation, has already descended—not touching the earth or any perishable substance but entering the diamond of the supreme prophetic soul. This golden Quranic sunlight radiated through Muhammad Mustafa during twenty-three years as a mercifully veiled revelation, as a life-giving sun that now circles and sustains the entire globe of human spirituality. This incomprehensible event—the vast sun plunging from the sky and entering a human heart—occurred on the Night of Power in the cave of silence on the mountain of exalted aspiration and yearning within the wilderness of willessness. That most precious human form, praiseworthy and trustworthy, was almost crushed in the revelatory embrace of Allah's Emissary of Light, until those most beautiful, bearded lips began to transmit the music of Ultimate Reality.

Our personal Night of Power, our own reception of the radiance and resonance of the Glorious Quran, comes every time we chant the Opening Sura, the ineffable Fatiha, disclosed by the sublime Oral Tradition to be equal in spiritual power to the entire Quran. Through this recitation, we mystically ascend through the boundless realms of revelation with seven steps of light—the seven verses of Sura Fatiha, which are the seven heavens and the seven levels of the self. Reaching the seventh mystic plane and sealing the confirmation of all revelation with the ecstatic cry *amin,* we can no longer be aware of our individual existence, for this would be idolatry, the elevation of created multiplicity to the level of uncreated Unity. Crushed in the all-powerful embrace of Sura Fatiha, permeated with its mystery, sparkling with its secrets, flowing like a healing spring with its Living Water, we enter the original Night of Power experienced by the holy Prophet, the Implanter of the Glorious Quran in the Heart of Humanity, may the infinite blessing of Sura Fatiha always be his.

Contained miraculously in Sura Fatiha is this mystic Night, which manifests outwardly only once a year, hidden by the Mercy of Allah in the last ten nights of Ramadan and illuminating each of these nights with its healing, awakening, shining embrace of blackness. Sura Fatiha is this Night of Power. Sura Fatiha is the entire Living Quran, and hence it contains the history and future of the universe. Sura Fatiha is the universal way of Islam revealed to humanity since the Prophet Adam, upon him be peace. Sura Fatiha is the purest Remembrance of Allah, the supreme *dhikr.* Sura Fatiha is the merging of the dervish soul into the Light of Reality. Sura Fatiha is the medicine for the illness of self-centeredness. Sura Fatiha is the channel of Divine Mercy that flows into every heart. Sura Fatiha is the Day of Judgment, the End of Time, and the Resurrection of manifest Being on the plane of Ultimate Reality, which is beyond Being. Sura Fatiha is the master key carried by the *walis* of Allah, hidden saints of the complete realization of Unity, and the *pirs* of the Tariqas, fountains of mystical teaching on the path to highest enlightenment.

Sura Fatiha, so indescribably powerful, is sung sweetly and simply in every prayer. Sura Fatiha is the prayer of Islam, the foundational prayer of the human heart. Without opening the channel of Sura Fatiha, no person of prayer enjoys the complete connection with the Ultimate Source, the perfect communion with Allah, which is Islam. The living Fatiha, the Fatiha who walked upon this earth blessing it with his perfectly guided steps, the holy Prophet, states in his sublime Oral Tradition, which is the commentary of his life upon the inner secrets of Sura Fatiha,

"Whoever does not recite Fatiha in his prayer, that prayer is incomplete." The supremely devoted Abu Hurayra, perfect mirror of the Oral Tradition of the Sun of Knowledge, proclaims, "If you recite Fatiha only, it is sufficient."

May our prayers and our very life breath be completed by Sura Fatiha. May our existence become the recitation of Fatiha only. Greetings, Holy Fatiha, Quranic Diamond at the heart of the Prophet of Allah, essential Message in the heart of the essential Messenger. Greetings, O flaming sun that gives light and life, self-veiling and self-revealing. May Sura Fatiha and the perfect human embodiment of these seven oft-repeated verses intercede for us now, as if it were the last moment of the universe. O Fatiha, show us the mystic resurrection. Awaken us to the Essence of Light through the seven levels of the self along your seven sacred steps. Lift us into your perpetual Night of Power and Peace, the Black Light, the Night of Union with the Beloved.

amin amin alhamdulillahi rabbi-l-alamin

SURA IKHLAS

What is sincerity, or *ikhlas,* other than living completely in the affirmation of the Unity of Reality? The sincere heart knows but a single love in union with Divine Love. The sincere mind knows only the single goal of union, affirming this goal instinctively and intuitively throughout the rich diversity of its thoughts and perceptions. The sincere mind never attempts to fragment Reality in its imagination, projecting false oppositions and dichotomies onto the perfectly harmonious and just expanse of Allah's Creation. The sincere heart perceives and cherishes only the existence of Allah, Who is Truth.

O Self-subsisting Truth, *ya haqq,* expressed completely through all beings yet abiding completely beyond all beings, please lift the sincere heart and mind into the timeless clarity and simplicity where the Divine Attribute of Truth becomes the actual spiritual state and station of the soul. Accomplish our dying into Truth and resurrection as Truth by the inconceivable power of Your Quranic Sura Ikhlas, O inconceivable Lord of Truth!

The four luminous verses of Sura Ikhlas, this organ of vision in the spiritual body of the Living Quran, are precisely the four mystic steps the dervish takes to approach the hand of the shaykh, which becomes the Hand of the Prophet, the Hand of Transcendence, the Hand of Truth, the Hand of Union. Each time the sincere lover of Truth recites the four verses of the Sura of Sincerity, these shining *ayats,* these Divine Signs, become the actual steps of *sharia, tariqa, haqiqa,* and *marifa*—sacred law, mystic path, union, and servanthood. The entire path of return into the Source is thus traversed in a few intense seconds of Quranic recitation, leading from the beautiful broad foundation, the noble Five Pillars of Islam, up long stairways of light, through chambers of intuition, mystical training, and teaching, to the roof garden of the palace of realization, entirely open to the sky of Truth, with no intervening form, structure, figure, pattern, concept, design, object, or subject. At the intonation *amin,* one awakens within Allah as the servant of Allah.

The sublime Oral Tradition affirms that this small Sura Ikhlas, although appearing just a brilliant drop of molten gold in the golden ocean of the Glorious Quran, actually contains one third of the creative and transformative power of the entire Book of Reality. Sura Ikhlas is the solace and joy of every sincere heart and mind who has awakened and responded to the invitation of Truth, embracing and being embraced by the universal way of Islam. This royal way dethrones the limited self, disclosing true humanity as the sultan of prayer and the crown of creation. Sura Ikhlas is the royal robe of the realization of all-embracing Oneness and the royal bearing of the lovers who experience their very existence to be the Self-subsistence of Truth. Sura Ikhlas is the Divine Instrument that creates mystic *faqirs,* those who ask for nothing other than Allah and who have lost every possession and every level of themselves in the boundless Light of Ahad, the Oneness Who imparts ineffable diversity to creation while subtly removing all diversity from sincere minds and hearts. The blessed Divine Name of Oneness, *ya ahad,* appears twice in Sura Ikhlas.

A soul once entered Medina the Illumined, sincerely intending before Allah to complete the Great Pilgrimage to Mecca the Ennobled. After taking full ablutions, body shining with the sacred lustrations in which water becomes liquid light and every part of the body it touches brightens with heavenly purity, this yearning soul, surrounded by beautiful dervish companions, entered the refreshment of sleep, mercifully ordained by Allah Most High. Since this soul came sincerely to the abode of the Prophet, it received the generous hospitality of the beloved one of Allah, whose presence has sanctified and sweetened the very dust of the valley of nearness called Medina. The most tenderhearted Muhammad, upon him be abundant peace, clearly perceived with his all-seeing eyes, which are simply the Eyes of Divine Mercy, this pilgrim soul, swimming joyfully in the companionship of true dervishes and surrounded by the powerful protection and intercession of a true mystic Shaykh, who had led this soul by the hand from the modern world of carelessness to the ancient desert of carefulness. As this holy Shaykh Muzaffer maintained nightlong vigil beside the radiant palace, the Tomb of the Prophet, the aspiring soul under his care was sent a dream from the highest Paradise. While dreaming, this fortunate soul was informed that it would be allowed to pose, from the depth of its being, a single question to the Shaykh of shaykhs, the beloved Prophet of Allah.

Not strong enough to bear a direct vision of the radiant Sun of Knowledge, the soul conveyed its question to the Prophet by intermediary—a question concerning certain passages in the Holy Quran implying that all beings in creation, regardless of their way of life or their level in the kingdoms of life, are bowing to Allah and praising Allah with every thought and action, however limited or sublime. After posing in the dream realm what was to be its ultimate question, this pilgrim was awakened suddenly, hair still wet from full ablutions taken before sleeping. Held in the gentle embrace of the Sacred City, surrounded by the perfect silence of the hours before dawn, the sincere soul was plunged into the depth of the mystery of Divine Unity. Its immense question had been asked in the light of the Holy Quran, and now the answer came through this clear Quranic Light, streaming directly from the diamond soul of the Distributor of the Light of Prophecy to the Hearts of Those Living in the Completion of Revelation. With the conscious power of one third of the Quran, the condensed Arabic verses of the noble Sura Ikhlas flowed into the whole being of the soul as the generous response of the Final Prophet to its own final question.

No distance remained between Quranic Sura and grateful recipient. Question and questioner were absorbed into the absolute sincerity of Oneness, like pools of rain drawn up by the noonday sun. Weeping copious tears beyond limited emotion, the soul took ablutions and performed two cycles of the Prayer of Islam, during which every movement was sincerity, every breath sincerity. Nothing existed at that moment other than the Sura of Sincerity, which proclaims and offers the experiential key to the highest Truth. Allah is complete Oneness. Nothing can exist outside all-embracing Oneness. Universe after universe within this Divine Creation are simply the Attributes of Allah praising the Essence of Allah. There is only One to pray and to receive the prayers. There is only Allah, only Allah, only Allah.

The sublime Oral Tradition of beloved Muhammad Mustafa, the Prophet Who Responds to the Final Question of Humanity with the Ultimate Disclosure of Unity, recounts a profound spiritual experience that occurred in Medina fourteen centuries ago. This experience continues to blossom today in the Medina of the sincere mind and heart. One of the lovers of Truth—a spiritually mature person, leader of the prayers among the Helpers of the Prophet, those who had welcomed him to Medina—used to recite only Sura Ikhlas, along with Sura Fatiha, through all the cycles of prayer, from dawn to midnight, day after day. The people complained to their Prophet about this repetition, asking for a greater variety of Quranic verses in the prayers. The beloved one of Allah, sublime servant and Messenger, addressed this lover of Sura Ikhlas:

> *What forbids you from doing what your companions ask you to do? Why do you recite Sura Ikhlas in every raqat?* The dervish lover replied with utter sincerity to this inquiry from his most exalted Shaykh, "I love this Sura." The Messenger of Allah replied, *Your love for Sura Ikhlas will cause you to enter Paradise.*

ESTAGHFIRULLAH

The constant cry of the dervish heart is *estaghfirullah, estaghfirullah, estaghfirullah*—may the healing stream of Allah's Forgiveness bathe the hearts of all human beings, cleansing every channel in their spiritual bodies from the subtle fire of the negation of love, the rebellion against love, the self-centered distortion of love.

At the thought of any action of its own that has fanned flames of negativity in others, injecting the venom of deception or the poison of division into their precious bodies of love, the dervish heart cries out for forgiveness—*estaghfirullah, estaghfirullah, estaghfirullah.* The dervish soul of love, hypersensitive to subtle forms of the rejection of love, always includes itself in its spontaneous prayer for Allah's Forgiveness of all humanity—*estaghfirullah, estaghfirullah, estaghfirullah.* The more it hungers and thirsts for righteousness, the more intensely does this soul pray the continuous prayer of longing for the uprightness and dignity of all human beings—*estaghfirullah, estaghfirullah, estaghfirullah.*

The dervish soul exists in conscious loving communion and community with all souls. Whenever the limited self attempts to separate from others, placing itself in opposition to others or in competition with others or making itself seem superior to others, the sensitive dervish heart cries out, May Allah forgive me!—*estaghfirullah, estaghfirullah, estaghfirullah.* Whenever the limited self attempts to claim credit for any holy or helping action, this sensitive heart immediately and instinctively, like a drowning person gasping for a breath of air, cries inwardly with great power, *estaghfirullah, estaghfirullah, estaghfirullah.* Whenever the limited self in its supreme foolishness blames the All-Merciful, All-Embracing Will of Allah Most High for its own negative actions or negative responses to any situation, the heart of purity, safeguarded by the holy way of submission, sincerely cries out, *estaghfirullah, estaghfirullah, estaghfirullah.*

The friends of Allah are being invited by their Lord along the mystical path. They are being exalted in humility. They have received clear signs of this sublime invitation and secret inward exaltation through taking the hand of a true shaykh, a mirror on earth of the Guide of Mystic Guides, the beloved Shaykh of Humanity, Muhammad the Messenger. With the living energy of Allah's Own Forgiveness, these courageous dervishes of self-forgetfulness cry out at the very claim of the limited self to have any rights of its own, *estaghfirullah, estaghfirullah, estaghfirullah.* At the threshold of spiritual perfection, where human life begins to become what Allah All-Generous created it to be, the dervish stream of constant Divine Remembrance even rejects the claim of the limited self to any existence whatsoever apart from the One Reality. Were this mature dervish soul to think in terms of *me* and *mine,* even for a second, its cry would blast forth like the trumpet at the End of Time—*estaghfirullah, estaghfirullah, estaghfirullah.* Even after the slightest distraction of awareness from the only Reality, such sensitive hearts feel the need to renew ablutions, repeating with genuine fervor *estaghfirullah, estaghfirullah, estaghfirullah.*

The ultimate cry for Divine Forgiveness, the *estaghfirullah* that repents of the very principle of separate self-awareness, the *estaghfirullah* that longs to merge in Allah and to live within Allah through the instrumentality of Allah alone, is the sweetest word that can pass through human lips. This perfect *estaghfirullah* is the resurrection into Paradise of all humanity without exception. This *estaghfirullah* is no longer a prayer but a spiritual station. This is the tear of love from a saint of love that extinguishes the fires of hell: *estaghfirullah, estaghfirullah, estaghfirullah.*

The most sublime prayer for forgiveness is reported in the noble Oral Tradition of the Prophet of Forgiveness, the One who Intercedes for the Forgiveness of Humanity at the Abode of Essence, the supremely tenderhearted Muhammad, may the Forgiveness of Allah flow through him into all the worlds, which were created through his Light and for his Light. Abu Hurayra, the intimate dervish lover of the holy Prophet who carefully observed every visible detail in the life of the Master of masters, noticed that during the prayer, the beloved one of Allah kept silent for a fleeting moment between the Words of Transcendence, *allahu akbar,* and the recitation of the Living Word of Allah, the Resonant and Radiant Quran. For the spiritual benefit of future generations, this intensely devoted companion, may Allah elevate him always into companionship with Divine Love, inquired of the Prophet of Love what he said to his Lord during that brief interval of intense inwardness and intimacy. The response illuminates the essence of *estaghfirullah,* the beauty of *estaghfirullah,* the wonderful intensity of *estaghfirullah.* These precious words, totally fresh after fourteen centuries of careful remembrance, cleanse the hearts of lovers and empty them entirely of the deceptions of the limited self. The supreme servant of Allah answered his devoted disciple immediately and openly about his own secret inward *estaghfirullah.* O brothers and sisters, listen to the sweet voice of the Sultan of Lovers that heals negativity, that puts out the hellfire of the negation of love, that reveals the heights of humility, and that washes the dust of limited desires and concepts from the robe of consciousness.

> I asked the Prophet, "What do you say in the pause between *takbir* and recitation?" The Prophet answered, *I pray: O Allah, set me*

> *apart from my faults as the East and West are set apart from each other. Cleanse me of faults as a white garment is washed clean. O Allah, wash away my faults with water, snow, and hail.*

Thus speaks the Culmination of the Light of Prophecy within the secret heart of humanity, which responds gratefully and ecstatically, May Allah forgive us!—*estaghfirullah, estaghfirullah, estaghfirullah.*

SALAWAT AND HADITH Compiled during a more recent Ramadan, this miniscule selection from the thousands of authentic Oral Traditions of the Prophet is a meditation in itself, without need for commentary. Arranged thematically, these pregnant sayings are each preceded by a different *salawat,* or prayerful invocation of the Messenger of Allah through his resonant names and radiant epithets. The resulting atmosphere of reverence tangibly enhances our capacity to receive the blessing of his words, spoken in the present tense of eternal companionship. These are not obscure Hadiths favored solely by mystics but are selected from the most popular and widely circulated anthologies. Expanded into dignified English, these inspiring utterances represent Muhammad the Messenger as the vast majority of Muslims have always understood him—a person of the greatest tenderness and clement wisdom. Some of these Oral Traditions are *hadith qudsi,* which manifest directly through the Voice of Allah. The others are spoken from the same Ultimate Source through the tender human voice of the Messenger. These renderings are on the level of *tafsir,* or interpretation.

6
Salawat and Hadith
Praise of the Messenger and His Inspired Words

ya khatam ul-mursalin wa khulafa ir-rashidin
O Seal of the Divine Messengers and his rightly guided representatives!

Practice of Universal Religion

While pointing at his noble heart three times, the glorious Seal of Messengers proclaims:
True religion is right here!

The Man of Prayer, upon him be peace, remarks:
Prayer is Light.

The beautiful Light of Guidance, upon him be peace, reveals:
The supreme gate to goodness is a person praying in the depths of the night.

The Shaykh of all shaykhs advises his blessed companions:
Always consult your own heart directly. Righteousness is whatever the inward heart is tranquil about. Wrongfulness is whatever causes the inward heart to waver.

The First Light and Universal Intellect, upon him be peace, relates these words directly from the Source of the Universe:
I manifest in the manner in which each conscious being expects Me to manifest.

The supreme Master of the spiritual path advises:

Live every moment in this limited world as if you were a traveler in a strange land.

The wise Counselor for all humanity without exception continues:

At evening, do not expect to be alive by morning. In the morning, do not assume that you will survive till evening.

The Messenger Muhammad, upon him be peace, relates these words directly from the exalted Creator:

O children of Adam, how often you complain about time, yet I alone am time.

Noble Ahmad the Praiseworthy, upon him be peace, instructs:

Follow any negative action directly with a good action. The negative one will be erased and annulled.

Proclaims the Guide of all guides, upon him be peace:

Never permit yourself to remain angry.

Advises the Reservoir of Prophetic Teaching, upon him be peace:

Either speak sincerely about goodness, or remain silent.

Our subtle Master points out to his companions:

Part of becoming an excellent practitioner of universal religion is to learn how to leave alone whatever is not your particular spiritual responsibility.

Advises the Friend of all souls, upon him be peace:

Abandon instantly whatever makes you doubtful, and embrace wholeheartedly whatever genuinely frees you from doubt.

The noble Bearer of the Glorious Quran recites these words directly from the Revealer of the Quran:

O My servants, I do not permit Myself a single act of compulsion, so neither is compulsion of any kind permissible among you.

Proclaims the nonviolent Warrior of Truth, upon him be peace:
Among you there should be neither harming nor reciprocating harm.

Proclaims the tender Intercessor who wept all night for the forgiveness of his spiritual community:
Never turn away from one another or undercut one another, even to the slightest degree.

Warns the Messenger of Universal Religion, upon him be peace:
A faithful person is brother to all. He never oppresses his spiritual brothers, never fails them, never deceives them, never regards them as inferior.

Pronounces the President of the Parliament of Prophets:
No person has become truly faithful to universal religion until he wishes for his brothers exactly what he wishes for himself.

The Final Prophet, upon him be peace, speaks clearly:
Of all that I have shown you to perform as universal religion, humbly perform as much as you can.

The noble Mahmud, upon him be peace, states with perfect clarity:
True religion is sincerity of heart!

The noble Mahbub, upon him be peace, confirms:
Purity of heart constitutes one half of universal religion.

The sublime Leader of the prayers for all humanity advises:
Worship Allah Most High with the intense sincerity you would feel if you directly perceived Him. You can attain this intensity by knowing full well that He is directly perceiving you.

The Bearer of the joyous news of Paradise remarks:
No person has become completely faithful to universal religion until his entire inward inclination is in perfect accord with me and with the knowledge I have brought.

The supreme Lover of Truth, who opens the gates of Paradise to all humanity, clearly affirms:

> *Whoever sincerely follows any path to seek spiritual knowledge, Allah Most High will open and make easy for him the way to Paradise.*

Reports the nonviolent Warrior of Divine Love, upon him be peace:

> *I have been ordered and ordained by Allah Most High to wage spiritual warfare with my people until they clearly witness that there is no reality apart from Allah and that Muhammad is a Messenger of Allah, until they perform pure prayer and give selflessly in charity. Then only will they have gained true protection.*

The Intercessor for all spiritual nations, for subtle beings, and for angels, upon him be peace, relates these words directly from Supreme Reality:

> *If My devoted servant draws near to Me by the measure of a hand, I draw near to him by the measure of an arm. If My humble servant approaches Me at the speed of walking, I come to him at the speed of running.*

Charitable Giving

Comments the Beloved One of Allah, upon him be peace:

> *Loving generosity is alone the true demonstration of universal religion.*

Declares the noble Prophet of the Arabs:

> *Each part of the physical body should perform some kind of charitable giving every day.*

The Bearer of the Universal Message, upon him be peace, remarks:

> *Charitable giving extinguishes negativity as effectively as water extinguishes fire.*

The Prophet of Love, upon him be peace, explains to his companions:

To mediate justly between two people is charitable giving. To help a person with his mount or his baggage is charitable giving. To offer a single good or kind word is to give generously in charity. Every step taken on the way to prayer is pure charity to the world. Just to remove a sharp or dangerous object from a path is to give in charity to all humanity.

The intimate Companion of Archangel Gabriel, upon them both be peace, reveals:

Whosoever helps any person in any form of pressing need, Allah Most High will immediately change his destiny, both here on earth and in eternity.

Confirms the Friend of humanity, upon him be peace:

Allah Most High will offer special assistance to His humble servant as long as that servant continues consistently and kindly to assist his brothers.

Reveals the Mercy to all Worlds, upon him be peace:

Every single repetition of praise—every subhanallah, every alhamdulillah, every allahu akbar, and every la ilaha illallah—is a supreme form of charitable giving.

Confirms the beloved Mustafa, upon him be peace:

Offering all praises to Allah alone with the sublime affirmation alhamdulillah entirely fills the Divine Scales of Justice.

The beloved Mujtaba, upon him be peace, relates these exalted words directly from Allah Most High:

O children of Adam, pour forth whatever wealth has been given to you upon anyone in any kind of need, and I will pour forth My Divine Wealth upon you.

The one who is intimate with the Lord of the Worlds relates these tender words directly from the Most Merciful:

I will divinely proclaim on the awesome Day of Resurrection, "O children of Adam, I was ill and you refused to nurse Me." They will reply, astonished, "How could we nurse You Who are the Lord of the Worlds?" I will then respond individually to each soul, mentioning the name of someone whom it knew on earth, "Do you not remember when My servant fell ill and you failed to visit him? If you had come to nurse this person, you would have found Me through him. Precisely the same is true about all those you have not fed or given drink."

Infinite Divine Forgiveness

The Mercy of Allah to all Worlds relates these words directly from the exalted Creator of the Worlds:

O children of Adam, so long as you call directly upon Me and supplicate with intense sincerity, I will always forgive you whatever you may have done and will take absolutely no offense. O children of Adam, were you to come before Me with negative thoughts and actions as vast as the entire earth—facing My Unity alone, without ascribing any duality to Supreme Reality—I would present you an equally vast forgiveness.

The noble Caretaker who guards the ocean of Quranic meaning, Prophet Muhammad, upon him be peace, proclaims:

When Allah Most High decreed the existence of creation, He firmly committed Himself by inscribing in His Original Book, "My Divine Mercy extends infinitely beyond My Divine Judgment."

Explains the ecstatic Lover of Allah, who prays constantly, "O Lord, increase me in knowledge":

Actions consist solely of the energy formed by the intentions behind them. Each person will receive from Allah solely upon the basis of what he intends.

The Messenger of the infinite Divine Mercy, upon him be peace, explains carefully to his blessed companions:

Concerning the person who sincerely intends a good action but fails to complete it, Allah Most Merciful records this intention in the Book of Divine Awareness as a completed good deed.

Concerning the person who both intends and completes a good action, Allah Most Merciful records it as ten good deeds or increases its spiritual value by as much as seven hundred times or even more.

Concerning the person who intends a negative action but does not actually perform it, Allah Most Merciful regards and records this very omission as a full good deed.

Concerning the person who both intends and actually performs a negative action, Allah Most Merciful records it as only one bad deed, or forgives it entirely.

Reveals the Beloved One of Allah, upon him be peace:

Out of love for me, Allah Most High has miraculously pardoned whoever may join my spiritual community for all their inadvertence, forgetfulness, and negative actions performed under various external pressures.

The Messenger of Allah relates these Divine Words directly from Allah:

Rebellious persons who sincerely feel My Divine Attraction and merely sit near those who are immersed in praising Me will be forgiven all their rebellious actions.

The spiritual Sovereign dressed in the garb of a humble man relates these Divine Words directly from the Lord of Love:

O My servants, without even being aware, you are constantly falling into negative thoughts and actions, day and night. But seek forgiveness directly from Me, and your negations will be erased and annulled.

The Prophet of Allah, upon him be peace, relates these words directly from the Lord of Power:

Whoever dares to claim, or even swears by Me, that I will not forgive a certain one of My servants, will thereby have his own good actions nullified.

Relates the Distributor of the Light of Prophecy to all hearts, upon him be peace:

A certain rebellious man directed his sons to cremate his body when he died. Brought before the Most High, this soul was asked why he had made such a request. He replied that it was solely from fear of meeting directly with Allah. For this alone, the All-Merciful forgave the man every one of his many offenses.

The mystical Traveler who ascended through the seven heavens into the Direct Presence of Essence reveals:

During the final third of the night until the first light of dawn, Allah is ceaselessly calling, "Who is praying that I may respond? Who is supplicating that I may grant? Who is seeking forgiveness that I may forgive?"

The Lover of Humankind whose intercession is always successful relates this mystery to his blessed companions:

When I encounter my precious Lord on the Day of Resurrection, I will fall in prostration and remain there as long as it may please Him. Thereupon, He will divinely proclaim, "Raise your head. Intercede, and your intercession will be accepted." I will begin to praise my Lord with a unique form of praise that He will teach me at that moment, and then I will intercede four times for all human beings to enter Paradise. During the fourth intercession, there will emerge liberated from the fires of hell persons who have affirmed la ilaha illallah only once and who possess only a single atom of goodness and kindness in their hearts.

The Gatherer of all humanity under the green banner of praise relates these words directly from Allah Most High:

O My servants, let whoever among you perceives goodness everywhere praise only Allah. Whoever perceives anything other than Divine Goodness can blame no one but himself.

Divine Presence

The beautiful Light of Guidance, *nur muhammad,* upon him be peace, transmits these words directly from the Most High:

> *O My humble servants, no being receives guidance except from Me. No being receives nourishment except from Me. No being receives protection except from Me.*

The First Light of Eternity, the Muhammadan Light, speaks these words directly from the Source of Light:

> *O My servants, you will never be able to harm or to benefit Me by your thoughts or your actions. No amount of devotion or worship can increase the sublimity of My Kingdom. No amount of negation or rebellion can decrease the sublimity of My Kingdom. Were I to offer every conscious being everything it requested, this could not diminish the abundance of My Kingdom, any more than dipping a needle into the ocean diminishes its depth.*

The intimate Beloved of Allah reveals these astonishing words directly from his most precious Lord:

> *When I intensely love one of My servants, I alone become the hearing with which he hears, the seeing with which he sees, the hands with which he grasps, and the feet with which he walks. Were such a servant to request absolutely anything of Me, I would instantly grant it.*

The beloved Messenger to whom Allah Most High addresses the words "But for you, I would not have created the universe" reports this promise directly from the Lord of the Universe:

> *Whenever they mention Me with love, I am with My devoted servants completely.*

The Light whose earthly crystal remains a fountain of radiance in Medina the Illumined reveals:

> *No persons can gather reverently in any House of Allah—*

reciting, studying, and meditating sincerely upon the Book of Allah—without infinite Peace descending upon them, infinite Mercy enveloping them, ranks of radiant angels surrounding them, and Allah Most High mentioning them with love to the beautiful souls who are nearest Him.

The Prophet of Love transmits these Divine Words that flow directly from the Source of Love:

If I love one of My servants intensely, I summon Archangel Gabriel and say, "I love this servant, so you too must love and support him." Then the Archangel speeds throughout the heavenly realms, crying, "Allah Most High loves this servant, so all of you must love and support him." Thereafter, profound love for this particular servant of Allah is gradually established among the beings on earth.

The Overseer of the tumult of universal resurrection, peace be upon him, relates these words directly from Allah:

On the Day of Resurrection, I will call out, "Where are those who love one another through My Divine Glory alone and for the sake of My Glory alone? Today, I am offering them refuge and sweet refreshment in My Shade, for this is a day on which there is no shade but My Shade."

The Essence of Paradise, the beloved Muhammad, upon him be peace, reveals these joyous words directly from his Lord:

I have prepared for My loving servants what absolutely no created eyes have seen, what no ears have heard, and what has never been imagined by any heart. So recite with true fervor the noble Quranic verse, "No soul can know what spiritual delight has been secretly prepared as the Divine Response to its earthly life of goodness."

The Sultan of Lovers, upon him be peace, reveals this ultimate mystery:

The Most High will ask the persons of Paradise, "Are you content?" They will reply, "Yes, Lord, for You have granted us what has not been granted to any other of Your countless creatures." The Most High will then continue, "Would you like to experience an even more exalted spiritual station? I will cause My Own Good Pleasure to descend upon you and abide within you. I shall never be displeased by your slightest thought or action."

The Divine Messenger who inaugurates the completion of revelation, upon him be peace, states with certainty:

Now the Divine Pen has been lifted and the pages have dried.

Brothers and sisters, let us chant together
the Essence of the Quran, the Mother of the Book,
the noble Sura Fatiha.

LIGHTNING FLASHES *Tasting* is a mystical term synonymous with *unveiling*. The clarified honey of prophecy must be tasted in our own mouths. Divine Light must be nakedly perceived by the eyes of the heart. These particular verses, in some cases fragments of verses, were favored by the consummate Shaykh, Ibn Arabi. The present English versions reflect his unveiling, or tasting, of the original revelatory Arabic. His full expansion of their meanings can be found in the marvelous book by William Chittick, *The Sufi Path of Knowledge.* Attempting to give some overview of the Quranic vision, I have arranged these verses thematically. The selections are called *lightning flashes* because they open beautiful dark clouds of essential meaning, drenching the intellect with life-giving rain. Understood intellectually, these verses remain just brilliant flashes, like heat lightning that does not bring rain. To be fruitful, they must occur within the clouds that gather in the sky of the heart during the storm of love. These renderings are on the level of *tafsir,* or interpretation.

7
Lightning Flashes
Verses Favored by the People of Tasting

Prophet Muhammad

Those who harmonize their whole being with the Messenger of Allah are harmonizing and unifying with Allah. *Quran 4.80*

The Messenger of Allah is the most beautiful pattern and empowerment for the holy way of life. *Quran 33.21*

Proclaim, O beloved Muhammad, "I am calling everyone to Allah through clear insight and direct experience, and so will those who follow in my spiritual lineage." *Quran 12.108*

Pray with your entire being, My beloved, "O precious Lord, increase and advance me in spiritual states and mystic knowledge."

Quran 20.114

Spiritual Guides

One of Our true spiritual servants, to whom We have granted Divine Mercy directly and taught Divine Knowledge mystically...

Quran 18.65

We divinely raise whomever We please to higher and higher degrees of wisdom. *Quran 12.76*

The Prophets and gnostics received Allah's direct inner guidance. Renew within yourselves the guidance they received.

Quran 6.90

Those who are always transcending, who are the foremost lovers, abide intimately near to Allah within the blissful Garden of Essence.

Quran 56.10–12

Their soul's light will race before them, emerging from the right side of their being.

Quran 66.8

Rivers of mystic wine, sheer bliss for those who drink there…

Quran 47.15

Those who mystically bear the Divine Throne and those who circle it in glorious praises to their Lord are completely immersed in Allah, crying out for the Divine Forgiveness of all beings.

Quran 40.7

The Nature of Allah

Every day and every instant, Allah shines forth as new spiritual states and new cosmic creations.

Quran 55.29

Allah warns you about the inconceivability of His Essence.

Quran 3.28

Nothing at all resembles Allah, yet Allah alone is the One Who is Seeing and Hearing.

Quran 42.11

Allah is the sole illumination of heavenly planes and earthly realms.

Quran 24.35

Behold how all beings and events are constantly returning home to Allah.

Quran 42.53

Allah is First and Last, Immanent and Transcendent, Manifest and Unmanifest.

Quran 57.3

The All-Knowing Mercy of Allah

Allah is ceaselessly All-Pardoning and absolutely All-Forgiving.

Quran 4.99

Allah's Mercy embraces every being and event without exception.

Quran 7.156

Allah will completely transform all negativity into goodness.

Quran 25.70

There is no conscious being who does not continuously praise the Divine Glory. Most persons simply cannot recognize or comprehend the universal praise directed toward Allah.

Quran 17.44

Allah alone creates you moment by moment, including whatever you think and whatever you do.

Quran 37.96

Not a single leaf falls without Allah's Knowledge, Permission, and Empowerment.

Quran 6.59

O Lord, You alone embrace all beings and events in Mercy and Knowledge.

Quran 40.7

The Spiritual Path

Remain constantly in profound awe of Allah, and Allah will teach you directly.

Quran 2.282

Give glorious praises to Allah, Who absolutely transcends what any finite being can conceive or describe.

Quran 23.91

The Ultimate Source now speaking is nearer to the human being than its own central vein.

Quran 50.16

In whatever direction or dimension you may turn, there is only the essential Face of Allah.

Quran 2.115

Affirm with your whole being, "Reality is now fully manifest and unreality has totally vanished, for unreality by its very nature must vanish."

Quran 17.81

Everything is annihilated except Allah's essential Face.

Quran 28.88

When My servants ask you, O beloved, concerning My Reality, I am already intimately near to them. I listen and respond instantly to the prayerful call of whoever calls.

Quran 2.186

Proclaim, O beloved, "Call upon the Only Reality as *allah* or *rahman* or with any of the beautiful Divine Names revealed to humanity, for all these Names belong to One Reality alone."

Quran 17.110

Keep prayerful vigil late into the night as an additional discipline, and Allah may raise you to the sublime station of constant praise.

Quran 17.79

There is no one who does not have a divinely appointed spiritual station.

Quran 37.164

The Process of Revelation

Our only Divine Speech to any being or event We will to create is the Word of Power *Be!* and it spontaneously comes to be.

Quran 36.82

You do not throw when you throw, but Allah alone throws.

Quran 8.17

Allah taught Adam the Divine Names as the fundamental nature of creation.

Quran 2.31

Allah has created humanity as the being with the most sublime spiritual stature in creation.

Quran 95.4

Allah, Who is Infinite Mercy, has generated the Cosmic Quran, has created an exalted humanity, and has revealed all possible meaning to purified human intelligence.

Quran 55.1–4

Proclaim that your Lord is Most Beautiful, for He wrote upon your heart with the Mystic Pen, revealing to humanity what it never could have known.

Quran 96.3–5

Every being in existence is revealed and contained in the Cosmic Quran.

Quran 6.38

Allah reveals Divine Signs to humanity both on the horizons of the universe and within the inner regions of heart, mind, and soul, so the realization will dawn that Allah alone is Real.

Quran 41.53

With Allah alone are the Keys of the Unseen, the treasures that no one knows but Allah.

Quran 6.59

There is nothing and no one whose inexhaustible treasures are not hidden within Allah.

Quran 15.21

Covenant, Mystic Vision, End of Time, Resurrection

Am I not your Lord? Yes, we witness to it.

Quran 7.172

O precious Lord, reveal Yourself to me that I may gaze directly into Your essential Face.

Quran 7.143

You will perceive the vast mountains that you imagined to be permanent dissipate like clouds in the open sky.

Quran 27.88

The entire planetary plane will shine with the ineffable Glory of its Creator and beloved Lord.

Quran 39.69

THREE DAYS OF PRAYER This selection of noble Quranic verses emerged from the communal practice of Ramadan during its intensified final ten days. A child born in the Holy Month is considered to be specially blessed; so it is with a spiritual child of Divine Light such as the present composition. This is no mere compilation. Each time the Glorious Quran is chanted, it is revealed anew by Allah Most High.

There are two poles of the Islamic week. Friday, the Day of Allah, begins Thursday at sunset and contains the universal congregational noon prayer, or *juma*. This gathering of Muslims across the planet provides a mystical reflection within time of the End of Time, when lovers of Truth will gather under the green banner of praise belonging to Muhammad the Messenger, upon him be peace. The second pole of the week, Monday, which ends at sunset, is the day of the Prophet's birth.

I selected these 120 verses to mark forty cycles of prostration each day during the three days of the week that contain both its spiritual poles, which are like the foci of an ellipse. Certain constellations of Quranic themes are indicated here by titles, yet the full range of Quranic vision is condensed during these fifteen periods of prayer. The verses are not meant to be studied but prayed. These renderings are on the level of *tafsir*, or interpretation.

8
Three Days of Prayer
Thursday Sunset to Monday Afternoon

We will always make this supremely simple and direct spiritual way joyously easy for you.

Holy Quran 87.8

sura fatiha By our ceaselessly invoking the Divine Name, Allah Most High—Who is tenderly compassionate, infinitely merciful—may perfect praise flow to Allah alone, Lover and Sustainer of all Worlds. He is most intimately called *rahman* and *rahim*. He presides magnificently over the Day of Divine Evaluation. O Lord, we worship only You and rely upon You alone. Reveal Your Direct Path, the mystic way of those who have received and truly assimilated Your sublime Guidance, those who never wander from the spiritual path and therefore never experience Your awesome Correction.

Quran 1.1–7

ayat al-kursi Allah alone. There is no divinity or reality apart from Him, the Living One, absolutely Self-manifesting and Timeless. His Divine Awareness never sleeps. All heavenly and earthly planes of being belong solely to Him. Only those whom He mysteriously permits can intercede before Him. He alone knows past and future. No one can grasp any of this panoramic Divine Knowledge except as He wills. The mysterious Throne of His Presence extends over every heavenly and earthly dimension as He effortlessly creates, protects, and preserves all manifestation. He is the Most High, the supremely Glorious.

Quran 2.255

asma al-husna There is no awareness apart from Allah, the all-encompassing Divine Awareness Who knows whatever is secret or open and Who is infinitely gracious and merciful. There is no divinity apart from Allah Most High—Divine Sovereignty, Divine Holiness, Divine Peace, Divine Faithfulness, Divine Protection, Divine Magnificence, Supreme and Sole Reality. Allah is incomparably glorious, free from any division or duality ascribed by the conventional mind. Allah alone is Creator, Evolver, Diversifier. To Him alone belong all the beautiful Divine Names. Simply by existing, whatever exists in heavenly and earthly realms declares perfect praise of His Gloriousness. He alone is supremely exalted, supremely wise.

Quran 59.22–24

THURSDAY SUNSET PRAYERS
Entering Paradise

Reclining on green cushions, seated upon rich carpets, surrounded by the most intense Divine Beauty, how could anyone ignore or deny the infinite generosity of the Lord?

Quran 55.76–77

For your beautiful lifetime of goodness, can there be any less reward than sheer Divine Goodness? How could anyone ignore or deny the infinite generosity of the Lord?

Quran 55.60–61

You will recognize upon their clear faces the magnificent splendor of Divine Bliss.

Quran 83.24

Those who cultivate profound faith, constantly demonstrating righteousness and sanctity, are true companions in the mystic Garden where they timelessly abide.

Quran 2.82

For them is reserved the Abode of Peace in the direct Presence of their Lord, Who will manifest as their most intimate and beloved Friend because they lived truly.

Quran 6.127

THURSDAY NIGHT PRAYERS

Exploring Paradise

The Night of Power

Those who are ever transcending, who are foremost among ecstatic lovers, abide in the blissful Garden of Essence, intimate nearness with Allah.

Quran 56.10–12

Behold the radiant Assembly of Truth, meeting timelessly in the direct Presence of the omnipotent Sultan.

Quran 54.55

Among these supreme lovers is passed a holy cup from the fountain of Divine Clarity.

Quran 37.45

Souls here call for and spontaneously receive every kind of nourishing spiritual fruit, as they abide in perfect peace, integrity, and sanctity.

Quran 44.55

We first divinely revealed luminous verses from this Glorious Quran during the most sanctified night, for We long eternally to offer Our compassionate Divine Warning. On that mystic Night outside time, every facet of infinite Divine Wisdom is made clear.

Quran 44.3–4

We began to unveil Our universal message during the Night of Power. How can its potency be described or conceived? This single Night is spiritually more intense than a thousand months, than an entire life of prayer. During this hidden Night, by the mysterious permission of their Lord, angelic guides and the Archangel of Revelation descend freely into all hearts. Divine Peace completely pervades creation until dawn.

Quran 97.1–5

We shall remove entirely from the blessed hearts of the lovers every conscious or unconscious sense of rancor or injury, and they will be spiritual compatriots in Paradise, facing each other timelessly on thrones of ecstasy, untouched by any sense of limitation or instability.

Quran 15.47–48

The only song in the Garden of Bliss will be "Glory to Thee, Allah Most High." The only greeting here: "May Divine Peace be upon you." The constant prayer here: "May perfect praise be offered eternally to Allah alone, Lover and Sustainer of all Worlds."

Quran 10.10

Some persons irrationally prefer the fleeting life of this world, but Paradise is inconceivably more excellent and absolutely enduring.

Quran 87.16–17

To the sanctified soul, Allah Most High communicates intimately, "My beloved soul who has attained perfect peace and conscious fullness, return now mystically into your Lord. Joining your spiritual delight with My infinite Divine Delight, enter now into the circle of My most exalted lovers, enter now into the Paradise of My very Essence."

Quran 89.27–30

To Allah alone belong East and West. In whatever direction or dimension you may turn, there is only the essential Face of Allah, Who is All-Encompassing and All-Knowing.

Quran 2.115

Those who constantly cherish only intense longing to encounter the essential Face of their Lord will attain complete realization.

Quran 92.20–21

Your most precious Lord is sheer Oneness. There is absolutely no reality apart from Allah Most High, Who is ceaselessly generous, infinitely merciful.

Quran 2.163

FRIDAY MORNING PRAYERS
The Allness of Allah

alif lam mim Allah alone. There is absolutely no reality apart from the One and Only Reality—the Living One, the Self-manifesting One, the Timeless One.

Quran 3.1–2

Do not call out to any apparent power apart from Allah, for there is only One. All existence is annihilated except Allah's essential Face. His alone is the Divine Command *Be!* To Him alone, all Being ceaselessly returns home.

Quran 28.88

All that exists upon earth or anywhere within time will perish, while the essential Face of your Lord abides timelessly, radiant with Majesty. How could anyone ignore or deny the infinite generosity of the Lord?

Quran 55.26–28

Affirm with your whole being, "True Reality is now completely manifest and all that is false and unreal has vanished utterly, for unreality by its very nature must vanish."

Quran 17.81

FRIDAY NOON PRAYERS
Mystic Gathering
Seeking Refuge beneath the Shade of the Holy Quran

alif lam mim This is the universal Book of Reality, the Divine Guidance, the Science of Certainty open to those who live in total awe of Allah Most High, who constantly acknowledge the Divine Mystery, who are ceaselessly engaged in prayer and acts of generosity, freely sharing whatever abundance We have divinely provided for them. These lovers respond wholeheartedly to the revelation streaming through you, My beloved, which is precisely the same message that flowed through all My previous Messengers. The persons who in this way assimilate Divine

Guidance directly from their Lord are the fully realized ones, the true human beings.

Quran 2.1–5

alif lam ra These are the powerful revelatory verses of the all-clarifying Universal Book. We have divinely sent it down in the form of a noble Arabic Quran so that you may easily learn mystic knowledge. Here is the most beautiful section of the Holy Quran. Before knowing this, you did not yet truly know.

Quran 12.1–3

O supremely compassionate Rahman, You ceaselessly transmit this Glorious Quran and create this exalted human form, revealing all possible meaning to purified human intelligence.

Quran 55.1–4

alif lam ra We have miraculously revealed Our Clear Book to lead humanity from deep darkness into sheer Divine Light. By sole permission of their Lord may persons enter the mystic way of the One, Who is infinitely powerful and Who is worthy of infinite praise.

Quran 14.1

Affirm with your whole being, "He is Allah, the One and Only Reality, the All-Encompassing, Who does not originate from some other reality. Nor do multiple realities come forth separate from Him, for apart from Him, there is absolutely nothing."

Quran 112.1–4

Affirm with your whole being, "I seek true refuge with the Lord and Lover of humankind, the mystic Sovereign of humankind, the Divine Guide of humankind—sure refuge from all negativity that whispers divisively into the pure heart of humanity and into the hearts of conscious beings on the subtle planes of being."

Quran 114.1–6

Whenever this Holy Quran is being prayerfully recited, listen to it with the complete attention of your entire being, keeping profound inward silence, that you may consciously receive the inconceivable Divine Mercy.

Quran 7.204

We have with Our infinite Grace bestowed upon you seven continuously repeated mystic verses and the vast and clear Quran that they subtly contain.

Quran 15.87

This Arabic Quran emanates mysteriously from the universal Mother of the Book, containing Our full Divine Presence, supremely exalted and full of spiritual wisdom.

Quran 43.4

This is a miraculous manifestation through Arabic letters of the mysterious Universal Quran, which remains timelessly inscribed upon a boundless transcendental tablet.

Quran 85.21–22

We have made this Arabic Quran easy to understand and remember. Are there any true human beings who long to receive its perfect guidance?

Quran 54.32

This is a perfectly pure Arabic Quran, free from the slightest human distortion, which provides most powerful protection against all negativity.

Quran 39.28

This is in Truth the sublime and noble Quran, most worthy of spiritual honor, the hidden Book of Reality that none can interpret deeply except those who are purified and sanctified.

Quran 56.77–79

Allah Most High erases or confirms whatever manifestation He wishes, for with Him is the mysterious Mother of the Book.

Quran 13.39

nun By the mysterious Divine Pen and by whatever it inscribes and reveals...

Quran 68.1

Proclaim with great power that your Lord is infinitely generous, revealing and teaching with the Divine Pen profound mysteries that the human mind could never discover.

Quran 96.3–5

FRIDAY AFTERNOON PRAYERS

Tolerant Way of the Holy Friends of Allah

Allah bestows mystic wisdom upon whomever He pleases, and such persons receive overwhelming blessings. No one can truly grasp My Message except these persons of profound insight.

Quran 2.269

Those who truly listen to My Holy Word and who actualize its highest level of meaning are the ones whom Allah guides by endowing them with deep understanding.

Quran 39.18

To all persons belong precise spiritual degrees, ranks, and stations that accord with the quality of their lives, for their Lord is never unmindful of what they think or do.

Quran 6.132

Pray with your whole being, "O precious Lord, bestow profound wisdom upon me and unite me with those who are intensely righteous and holy."

Quran 26.83

Never dispute aggressively with various Peoples of Revelation who remain nonviolent and sincere, but lovingly affirm to them, "We believe in the integrity of the essential Divine Revelation, which has come down to us as well as to you. Allah Most High is the One Reality to Whom we all bow and submit our souls."

Quran 29.46

There can be absolutely no compulsion in universal religion. Simply allow living Truth to stand out clearly from error. Whoever avoids all negation and lives by affirming Allah alone is one who has grasped the indestructible essence. Allah Most High is the Sole Awareness Who constitutes all seeing and all hearing.

Quran 2.256

If it had been Our Divine Will, everyone on earth would sincerely and openly believe. So why should you attempt to persuade or compel any person to become a believer?

Quran 10.99

Simply pray unceasingly, give generously, and assimilate the Messenger's holy way of life completely. You will receive infinite Divine Mercy.

Quran 24.56

FRIDAY SUNSET PRAYERS

Mystic Manifestation of Prophet Muhammad through the Arabic Letters *ta ha* and *ya sin*

ta ha Only Allah. There is absolutely no reality other than Him. He alone manifests the energies of all the beautiful Divine Names.

Quran 20.1, 8

ya sin This Quran is the full manifestation of Divine Wisdom, witnessing clearly that you, My beloved one, are among the exalted Messengers who lead humanity on the Direct Path, bearing the purest revelation from Allah Most High, Who is All-Powerful and All-Merciful.

Quran 36.1–5

Allah alone is Reality. Whatever existence may be invoked apart from Him is mere empty falsehood. He is All-Encompassing, All-Transcending.

Quran 31.30

He is the Living. There is neither divinity nor life apart from Him. With ecstatic devotion, call upon Him alone. Glorious praises be to Allah Most High, Lord of all Worlds.

Quran 40.65

Never proclaim about those who die consciously in the mystic way of Allah that they are dead. In essence, they live. But this fact cannot be perceived with ordinary eyes.

Quran 2.154

FRIDAY NIGHT PRAYERS

Awakening through Trials and Tests to the Exalted Spiritual Stature of True Humanity

We offered Divine Responsibility to the heavenly and earthly planes and to the primordial mountains, but they all refused it, overwhelmed with awe. Only humanity, although outwardly lacking in mastery and knowledge, dared to accept this most profound responsibility.

Quran 33.72

By the holy Fig Tree, by the blessed Olive Tree, by the sacred Mountain Sinai, by the ennobled and spiritually secure City of Mecca, We have created humanity as the most exalted form in Our Divine Creation.

Quran 95.1–4

alif lam mim Do persons imagine that they will be accepted as authentic believers just by proclaiming "We believe" without being profoundly tested?

Quran 29.1–2

We bring intensive trials upon aspiring souls, carefully testing those who strive their utmost for spiritual advancement and who persevere in pure patience, so that We can divinely demonstrate their true level.

Quran 47.31

The ills will be healed and the difficulties resolved of all those who profoundly believe and who constantly demonstrate righteousness and sanctity, fully accepting this revelation sent down to My beloved Muhammad as purest Truth from their Lord.

Quran 47.2

How can the spiritual way of steep ascent ever be fully described? It manifests most concretely as freeing the enslaved, feeding the hungry, taking personal responsibility for the orphans and the homeless. The practitioners on this holy way profoundly believe, constantly demonstrate patience, ceaselessly engage in kind actions and compassionate thoughts. These are the mystic companions of the Divine Right Hand.

Quran 90.12–18

To that man or woman who truly demonstrates goodness and sanctity, We divinely unveil a new life that is totally good and pure, elevating such persons to the exalted level of their own highest ideals.

Quran 16.97

Replicate within your own consciousness the wonderful spiritual states of all those who have received Direct Guidance from Allah Most High. Proclaim, O My beloved, "No reward for this inconceivably precious knowledge do I request from any person. This is the open message to all humanity called Universal Islam."

Quran 6.90

No person will be able to keep any of this Divine Message in clear remembrance except as willed by Allah Most High. He is the Lord of Righteousness and the Lord of Forgiveness.

Quran 74.56

You are able to will only as Allah wills. He is the Lover of all Worlds.

Quran 81.29

Intensely pray additional prayers in the sacred hours after midnight, and your Lord may elevate you to the rare station of constant praise and perfect illumination.

Quran 17.79

Standing and prostrating, these sublime aspirants spend the entire night in ecstatic adoration of their Lord.

Quran 25.64

Weeping intensely, they fall upon their faces with ever-increasing humility.

Quran 17.109

SATURDAY MORNING PRAYERS

Manifestation of Beloved Jesus, Blessed Virgin Mary, and Prophet Zakariya, Father of John the Baptist

The infant Jesus proclaimed, "I am in Truth a humble servant of Allah Most High, Who has entrusted me with His Revelation, manifesting me as His Messenger."

Quran 19.30

Allah Most High taught the Messiah Jesus both the Divine Book and Divine Wisdom, both Torah and Gospel.

Quran 3.48

kaf ha ya ain sad This is the miraculous demonstration of the inconceivable Mercy of your Lord to His true servant, the noble Zakariya.

Quran 19.1–2

Relate now from My Book of Reality the sublime spiritual history of the noble Mary when she courageously withdrew from her earthly family to a mysterious chamber facing East.

Quran 19.16

SATURDAY NOON PRAYERS

Manifestation of the Noble Prophets Noah, Abraham, Jacob, David, Job, and Moses

Among those who devotedly followed Prophet Noah's exalted spiritual way was the noble Abraham.

Quran 37.83

The noble Prophet Jacob revealed, "I weep to Allah with the intense anguish of my love for the beloved Joseph, yet I have received from Allah Most High a mystic knowledge that you cannot know."

Quran 12.85

At dawn and twilight, We miraculously caused hills and mountains to declare Our praises in joyful unison with the noble Prophet David.

Quran 38.18

Strike with your foot precisely here, O noble Job. Cool water will flow forth miraculously, cleansing you and slaking your thirst.

Quran 38.42

They encountered one of Our true spiritual servants, upon whom We had bestowed Our complete Mercy, teaching him mystic knowledge directly from Our essential Presence.

Quran 18.65

A mysterious guide proclaimed to the Prophet Moses, "You will not be able to bear patiently with me, for how can you experience true patience concerning events about which you lack full knowledge?" Moses replied, "You will find me, if Allah wills, patient and obedient to your mystic teaching."

Quran 18.67–69

We divinely called the noble Moses from the right-hand side of the sacred Mountain Sinai and caused him to approach very near to Us.

Quran 19.52

In Truth, I alone am Allah. There is no divinity, holiness, or awareness apart from the infinite *I Am* that I am. So be conscious only of Me, establishing constant prayers to remember, celebrate, and praise Me.

Quran 20.14

Proclaim with your whole being, "My Lord places the mystic robe of Truth over His true servants. He alone possesses full knowledge of all that is hidden and mysterious."

Quran 34.48

Among those We have created are certain chosen persons who can direct others purely according to Divine Truth and Divine Justice.

Quran 7.181

SATURDAY AFTERNOON PRAYERS
Noble Abraham: Spiritual Father of Jews, Christians, and Muslims

Relate now from My Book of Reality the spiritual history of the noble Abraham, the sublime Prophet, the perfect man of Truth.

Quran 19.41

Who could be more advanced in the path of true religion than the one who submits his entire consciousness to Allah alone, constantly demonstrating goodness, and who thus sincerely follows the mystic way of Abraham the True? Allah Most High embraced the noble Abraham as His intimate spiritual friend.

Quran 4.125

This is the Divine Reasoning that We revealed to the noble Abraham in order to refute the spiritual errors of his people. We mystically advance and elevate whomever We will, degree after degree, for your Lord is infinite in wisdom and knowledge.

Quran 6.83

We then divinely proclaimed, "O fire, become cool and protect the precious life of the noble Abraham."

Quran 21.69

Through His inconceivable, immeasurable Grace, Allah Most High rewards whomever He wills at the level of their most beautiful intentions and actions, and even above this level.

Quran 24.38

As for those who radically repent by turning entirely toward Divine Unity and who demonstrate this by following the holy way of life, Allah Most High will completely transform their negativity into goodness. He is ceaselessly forgiving, infinitely merciful.

Quran 25.70

O My servants who have transgressed against the innate nobility of your own souls, never doubt the astonishing Mercy of Allah Most High, Who can forgive and erase every negation and distortion. He is ceaselessly forgiving, infinitely merciful.

Quran 39.53

Mystically gracious with profound subtlety is Allah to His servants. He pours forth material sustenance and spiritual abundance to whomever He pleases, for He possesses infinite power to manifest His Will.

Quran 42.19

SATURDAY SUNSET PRAYERS

Intercession by the Sublime Friends of Allah
Universal Prayer of Beloved Jesus

Those who mystically bear the Divine Throne and those who circle around it with most glorious praises to their Lord are totally immersed in Allah Most High, crying out spontaneously for the Divine Forgiveness of all conscious beings, "O precious Lord, You embrace all lives with Divine Mercy and Divine Knowledge. Forgive utterly those who turn in pure repentance and who sincerely follow the spiritual path. Liberate all humanity from the terrible Fire of Rebellion."

Quran 40.7

Prayed Jesus, noble son of the Virgin Mary, "O precious Lord, send from Paradise a mystical table magnificently spread, so that every human being without exception may celebrate the profound festival of the Ultimate Divine Sign, thus providing true spiritual sustenance for all humanity. Allah alone is the best and only Sustainer."

Quran 5.117

To Allah Most High, the noble Jesus is like the noble Adam, whom He created directly from dust, simply proclaiming unto him *Be!* and he came to be.

Quran 3.59

To any being or event that We have divinely willed to create, We simply proclaim the Word of Power *Be!* and it miraculously comes into being.

Quran 16.40

To generate an infinite creation that is perpetually new, moment by moment, does not pose the slightest difficulty for Allah Most High.

Quran 14.20

SATURDAY NIGHT PRAYERS

The End of Time

Functioning of the Universe as Pure Praise

Every conscious being will come singly before Allah Most High on the timeless Day of Evaluation.

Quran 19.95

Some faces on that timeless Day will be completely radiant, gazing only upon their sublime Lord.

Quran 75.22–23

You will perceive the vast mountains that you imagine to be permanent dissipate like clouds in the open sky. This End of Time is also part of the sublime creativity of Allah, Who harmonizes all events and knows all thoughts.

Quran 27.88

The entire planetary plane will resplend with the ineffable Divine Glory of its Creator and Lord. The Universal Quran will open. All the holy Prophets and their spiritual witnesses will manifest. Divine Justice will prevail, and no soul will be wronged.

Quran 39.69

On the Day beyond Time, no intercessor will avail except those for whom clear permission to intercede has been granted by Allah Most Gracious—those whose spiritual words are acceptable to Him.

Quran 20.109

That timeless Day will transparently manifest Absolute Reality, and those who truly long to return will return consciously into their Lord.

Quran 78.39

Absolute Reality. What is Absolute Reality? What will make possible the realization of Absolute Reality?

Quran 69.1–3

The heavenly realms will be thrown open as though they were merely doors, and the primordial mountains will vanish like dreams or illusions.

Quran 78.19–20

Whether temporal creatures or angels, all beings on the earthly and heavenly planes bow deeply and constantly before Allah Most High, because no being can be essentially arrogant before Pure Being. All living creatures revere their Lord, consciously or unconsciously performing all that He commands.

Quran 16.49–50

The seven heavenly planes, the earthly plane, and the countless conscious beings they contain are spontaneously witnessing and declaring the Divine Glory and Unity. Nothing exists without celebrating and praising Allah Most High by its very existence. Most persons cannot recognize and comprehend this universal praise and the constant Divine Response of perfect patience and boundless forgiveness.

Quran 17.44

Every conscious being within the heavenly and earthly dimensions of Being essentially seeks and knows only the Source of Being. Every moment, the Source shines forth with fresh splendor as the entire universe. How could anyone ignore or deny the infinite generosity of the Lord?

Quran 55.29–30

There is no being of any kind whose inexhaustible secret treasures are not hidden within Us. But We divinely manifest these treasures by finite degrees.

Quran 25.21

I have created human beings and subtle beings only so they can consciously and joyously praise Me.

Quran 51.56

SUNDAY MORNING PRAYERS

Mystery of Revelation

Throughout the heavenly realms and across the whole earth, Divine signs are clearly manifest for those who profoundly believe.

Quran 45.3

O humanity, for those who profoundly believe there has come spiritual instruction from your Lord and healing for the diseases of your hearts. There has come perfect Divine Guidance and boundless Divine Mercy.

Quran 10.57

There came running from the farthest corner of the city a mystic, who cried out, "O my people, obey the Messengers of Truth."

Quran 36.20

It would be irrational if I did not devotedly serve the One Who created me, the One to Whom every conscious being inevitably returns.

Quran 36.22

SUNDAY NOON PRAYERS
Communion with the All-Forgiving Divine Will

Allah alone creates you, moment by moment, including whatever fruitful results your mind and hands create.
Quran 37.96

No event can occur, externally or within your own heart, that is not already divinely decreed before We bring it into visible manifestation. This perfect foreordination is simple and effortless for Allah Most High.
Quran 57.22

We create all manifest Being in perfect proportion and harmony. Our Command is a single Divine Act, like the flash of an eye.
Quran 54.49–50

Remain in complete awe of the One Who bestows upon you, moment by moment, all that you know.
Quran 26.132

Proclaim with your whole being, "Call upon the One and Only Reality with the Names *allah* or *rahman* or with whatever Divine Names have been revealed, for to Him alone belong all the beautiful Names. Pray neither loudly nor in mere whispers but harmoniously."
Quran 17.110

O My beloved Messenger, proclaim to humanity, "If you truly love Allah, follow me intimately and Allah will love you intensely, forgiving and thoroughly erasing all your negativity. Allah Most High is ceaselessly forgiving and infinitely merciful."
Quran 3.31

All persons without exception can hope that Allah Most High will forgive, for Allah utterly obliterates the traces of rebellion and error by forgiving continuously.
Quran 4.99

He is the ceaselessly forgiving One, eternally overflowing with most tender Divine Love, the sublime Lord of the glorious Throne, Who manifests moment by moment all that He divinely intends.

Quran 85.14–16

May glorious praises flow to Allah alone, Who is supremely exalted, supremely transcendent—high above whatever can be asserted or thought.

Quran 17.43

Joyfully proclaim, "Glorious praises be to our Lord Most High. Truly has the full promise of the Lord now been fulfilled."

Quran 17.108

SUNDAY AFTERNOON PRAYERS
Advancing on the Way of Mystic Knowledge

Allah alone exists, exalted be He, the sublime Sovereign, the Only True Reality. There is absolutely no source of power or holiness other than He, Lord of the glorious Throne of Mystery.

Quran 23.116

Allah Most High reveals miraculous Divine Signs to humanity, both on the vast horizons of the universe and within the intimate regions of their own hearts, until it becomes perfectly clear that Allah alone is Reality, the sole Witness of all manifestation.

Quran 41.53

Those who have received and truly assimilated Divine Knowledge perceive directly that the revelation descending through you from your Lord is the living Truth that guides all souls to the mystic path of the Only One, Who alone is worthy of infinite praise.

Quran 34.6

O humanity, it is you that have the fundamental need for Allah, whereas Allah Most High is the sole Reality, free from any possible lack and worthy of infinite praise.

Quran 35.15

No person lacks a divinely appointed spiritual station. All are precisely ranged in ranks and degrees, for human beings have received the supreme responsibility to declare consciously the inconceivable Glory of Allah.

Quran 37.164–66

Paradise is even more subtly differentiated into ranks and gradations of spiritual realization than temporal existence and is infinitely greater in excellence.

Quran 17.21

Those holy friends of Allah Most High who are arranged by spiritual ranks and degrees become immensely powerful in dissipating negativity by proclaiming My Message of Unity: "Your precious Lord is sheer, all-encompassing Oneness."

Quran 37.1–4

They speak only as He speaks, acting exclusively through His direct Divine Command.

Quran 21.27

SUNDAY SUNSET PRAYERS
Divine Light Manifest as Prophet Muhammad

Time is coming to its end and human beings are wasting time—except for those who possess complete certainty about the universal religion of Unity, who engage constantly in acts of righteousness, and who encourage one another, with great patience and constancy, in the sublime way of Truth.

Quran 103.1–3

Allah alone is the single Light of Truth that reveals both heavenly and earthly planes of Being. Meditate upon Divine Light manifest as a mystic flame, protected by transparent glass that glistens like a star. This spiritual lamp, placed in a high prayer niche, is kindled from the oil of a transcendent tree, not found in East or West—an oil luminous by nature, needing no spark of material fire. This light of the soul shines forth within Divine Light. Through such profound meditations, Allah Most High guides whomever He wills into His most intimate Light. Allah is Omniscience.

Quran 24.35

Send forth your powerful spiritual gaze into the perfection of Divine Creation again and again, My beloved. It will return dazzled, amazed, enlightened.

Quran 67.4

The most beautiful pattern and empowerment for the holy way of life has been manifested through the Messenger of Allah for anyone whose sole hope rests in Allah Most High and His eternal Day.

Quran 33.21

Allah and His radiant angels send continuous blessings upon the sublime soul of the Prophet, O noble believers, so you as well should humbly and constantly request Divine Benediction for him, saluting him repeatedly with utmost love and respect.

Quran 33.56

SUNDAY NIGHT PRAYERS

Mystery of the Prophet Muhammad His Call and His Enlightenment

We have called forth from among your own people a noble Messenger, clearly demonstrating Our Divine Signs, sanctifying you and instructing you in scriptural revelation. He bears both profound ancient wisdom and new mystic knowledge.

Quran 2.151

Those who instinctively lower their voices in the presence of Allah's noble Messenger have hearts well proven by Allah to be full of holiness and devotion. Theirs is the complete Divine Forgiveness and the most sublime Divine Reward.

Quran 49.3

O beloved, whenever My humble servants question you sincerely concerning Me, I am intimately near to them. I listen carefully to the prayer of every sincere supplicant who calls upon Me with intensity. May all humanity also listen to My Call, experience Me directly, and walk in the True Way.

Quran 2.186

Those who harmonize their whole being with the Messenger are harmonizing and unifying with Allah Most High. If any persons resist or reject you, My beloved, remember that We have not sent you to monitor or control their actions.

Quran 4.80

Affirm with your whole being, "I seek true refuge with the Lord of perpetually dawning Wisdom—perfect refuge from all the negativity generated by limited conscious beings, with which they darken their consciousness, total refuge from the confusion, magic, and obsession they project."

Quran 113.1–5

To you, O beloved, We have divinely granted the mystic Fountain of Abundance. You are oriented solely toward your Lord in constant inward prayer and compassionate sacrificial service. Those who separate from you are condemned to the painful illusion of separation.

Quran 108.1–3

We have never sent any holy Messenger previous to you, O beloved, without the identical essential revelation: "There is absolutely no reality apart from Me, so remember Me ceaselessly."

Quran 21.25

All-Transcending is Allah Most High—the Mystic King, the Living Truth. Never slacken your constant receptivity to the Holy Quran until its sublime revelation is completely unveiled. Never cease praying, "O precious Lord, advance me in spiritual states and knowledge."

Quran 20.114

O you who practice deep contemplation, completely covered by your wearing cloth, stand for prayer during half the night, reciting Holy Quran in slow, rhythmic, harmonious tones. We shall send a tremendous Divine Message through you.

Quran 73.1–5

O you who are concealed beneath the prophetic cloak in profound meditation, arise now and deliver My most compassionate Warning: "Magnify your Lord and keep your garment stainless."

Quran 74.1–4

The noble Gabriel appeared above the vast horizon and approached closer until he manifested radiantly at a distance of two bow-lengths or even nearer. Then Allah Most High transmitted to His servant Muhammad, through the most intense inspiration, whatever He divinely wished to transmit.

Quran 53.7–10

The flowering tree near the Garden of Essence that marks the mystic frontier beyond which no limited being may pass was suddenly engulfed by Divine Radiance. The sublime gaze of Muhammad did not waver or turn even slightly away, and he merged into the Direct Presence of his Lord.

Quran 53.14–18

The Divine Good Pleasure descended upon the believers when they swore eternal loyalty to you beneath the mystic tree. Allah Most High knew what was in their hearts, so He sent down profound tranquility upon them and rewarded them with true victory.

Quran 48.18

MONDAY MORNING PRAYERS
Spiritual Victory of the Prophet Muhammad

We sent down this Quran in Truth, and with Truth alone has it descended. And We sent you, O beloved, to transmit its wonderful promise and its merciful warning.

Quran 17.105

O Prophet, We have sent you as a pure witness, as a bearer of joyous news and a compassionate warner, as a lamp shining with Divine Light—as one who, by Divine Permission, invites all persons without exception to the Truth.

Quran 33.45–46

Have We not vastly expanded your heart, O beloved, removing completely from you the burden of universal human suffering that you bore painfully upon your shoulders? Have We not revealed you as the human being most worthy of praise? With every intense spiritual struggle comes spiritual delight; with difficulty comes ease. Whenever you are free from external responsibilities, continue to pursue interior prayer, giving every moment of awareness to your precious Lord.

Quran 94.1–8

When miraculous Divine Help brings spiritual victory from Allah Most High, you will observe multitudes awakening into universal religion. So celebrate constantly the most glorious praise of your Lord, and pray fervently for His all-embracing Forgiveness, which flows forth again and again to all conscious beings.

Quran 110.1–3

MONDAY NOON PRAYERS
Divine Instructions to Beloved Muhammad

Affirm with your whole being, "My prayer and my sacrificial service, my entire life and my death, are dedicated solely to Allah Most High, Lover and Sustainer of all Worlds."

Quran 6.162

There has manifested through you from Allah Most High, O beloved, a Book of Clear Light and a new spiritual radiance.

Quran 5.17

O My noble Messenger, joyously sufficient unto you is Allah and unto those among the faithful who truly follow you.

Quran 8.64

My beloved, ask for no recompense whatsoever from any person for this inconceivably precious revelation, which is My universal Message for all creation.

Quran 12.104

Proclaim, O beloved, "It is my spiritual way simply to invite all persons without exception to Allah through gnostic insight and direct vision. This will be the spiritual way of all those in my lineage, who will give glory only to Allah Most High and who will never associate the slightest duality with Him."

Quran 12.108

O beloved, We have sent you forth for no reason other than to be Our Divine Mercy for all conscious beings.

Quran 21.107

O beloved, proclaim to all humanity, "I ask absolutely no reward from you except that you wholeheartedly take the direct path to your Lord."

Quran 25.57

O beloved, you will not be able to guide everyone you love, for Allah Most High guides whomever He wills, and He alone knows who is actually receiving Divine Guidance.

Quran 28.56

Only that practitioner who offers to Allah Most High a profoundly serene and balanced heart will be able to progress spiritually.

Quran 26.89

Affirm with your whole being, "It is Allah alone Whom I serve with total ecstatic devotion."

Quran 39.14

MONDAY AFTERNOON PRAYERS
The Fullness of Islam

Allah Most High has divinely decreed, "I and My Messenger must be victorious." And the Will of Allah is infinitely powerful.

Quran 58.21

Pay no attention to those who hold divisive views, but prostrate in pure adoration to the One Reality, becoming more and more intimate with Allah alone.

Quran 96.16

Place your complete trust in Allah alone, and be firmly established on the beautiful Path of Truth.

Quran 27.79

This is the Ultimate Truth, known with mystic certainty. So celebrate with infinite praises the exalted Name of your Lord, who is the Supreme and Sole Reality.

Quran 59.95–96

May we celebrate pure praises of You ceaselessly, enjoying the constant inward remembrance of True Divinity.

Quran 20.33–34

Truly celebrate the perfect praises of your Lord by being among those who spontaneously and continuously prostrate in adoration.

Quran 15.98

Such is the wonderful Generosity of Allah Most High that makes it sufficient for us to know that Allah alone knows all.

Quran 4.70

He alone is First and Last, Manifest and Hidden, Immanent and Transcendent. He alone possesses full knowledge of all Being.

Quran 57.3

THE GENEROSITY OF ALLAH Here is another spiritual child that came to birth within the Holy Month of fasting from limitation and feasting on limitless Truth alone. This composition explores the mystery of Ramadan through another mystery, Sura Rahman, the sweetest chapter in the entire Quran. The approach is that of *tasting*. The supreme secret of identity is adumbrated here. If Allah is infinitely generous, will He offer any less than an infinite realization to His beloved servants?

9
The Generosity of Allah
Ramadan and Sura Rahman

bismillah ir-rahman ir-rahim Only in the healing, illuminating, empowering, and sanctifying Name of Allah, Who is absolute Mercy and inconceivably tender Compassion, can we begin the infinite task of elucidating the mystery of the holy month of Ramadan.

ya wadud O Allah, You are none other than Love—a single love in countless beautiful forms and brilliant facets. Only through awakening as Your Divine Attributes, manifest as both lover and beloved, can we envision, approach, and merge with Your Reality, which is the Only Reality. Your undifferentiated Presence, O Allah, pours through the month of Ramadan like an iridescent waterfall, each drop containing the complete spectrum of Your ninety-nine essential Energies, from delicate Beauty to all-consuming Power.

Only through the open gates of the Paradise of Ramadan can the total flood of Divine Presence enter, awaken, and transform the heart. The fully flowered state of sanctity is hidden within a few human beings, but the noble month of Ramadan is open and available to persons at all levels of spiritual development. Ramadan reveals the intrinsic sainthood of every heart.

Wherever the human level of consciousness may manifest on suitable planets throughout galactic space, it is the jeweled crown of Your sublime Creativity, O Allah. It is consciousness become transparent to its own Source. True humanity consists of Your Attributes turning and gazing at Your Essence, crying out through their intrinsically Divine Being, *la ilaha illallah,* the One alone exists.

What is truly human is the return, through the precious human form, of the radiant Divine Energies into their Source, the single Essence. This return is universal *fana,* disappearance into Allah. Then there blossoms the selfless flowing forth of Divine Attributes again as perfect humanity, which is universal *baqa,* subsistence in Allah. The hands of this perfect human being are *ya rahman* and *ya rahim,* the continuous invocation of Divine Mercy and Compassion. The breath of this fully realized humanity, promised by the whole prophetic tradition, is *ya hayy,* the ceaseless calling out to Divine Life, which constitutes all lives. The heart of the awakened human being manifests only Divine Justice, the eyes see only Divine Beauty, the mind experiences only Divine Clarity. This person is the lover of humankind and the tender *khalifatullah,* or manifester of Allah, upon the planetary plane of existence—cherishing, protecting, and spiritually maturing all the creations of Allah as they unfold in universal balance and harmony.

This magnificent awakening of the human being as the mysterious inflowing and outflowing of Divine Essence occurs during the holy month of Ramadan with the ease experienced in Paradise. This most sacred month is the actual opening of the Gates of Eternity into the realm of time, inundating the cosmos of the heart with the rose fragrance of the Paradise heart of Prophet Muhammad, may supernal peace surround him. During this most mysterious month, the lovers of Allah can experience the same sweet taste in their mouths that the Holy Prophet experienced, the very honey of prophecy.

Ramadan is the most intense expression of Divine Generosity that humanity can encounter. It is the month during which the

Glorious Quran descended fourteen centuries ago. During Ramadan, the Quran actually descends afresh each year, as healing and illuminating Divine Presence. This miraculous descent is most clearly focused during the Night of Power, hidden among the odd-numbered nights in the last ten days of Ramadan. Consciously to enter this luminous Night is equal in intensity and spiritual growth to one thousand months of prayer, as Allah confirms in His Book of Reality.

Prophet Muhammad—upon him be the sublime peace of the mystical union to which he draws all humanity with the vast magnet of his love—is the complete Mercy of Allah, the Culmination of Prophecy, the return of humanity into its essential Source, the Distributor of the single Light of Prophecy to the hearts who enter Ramadan through the vast, open portal of the Clear Quran.

This diamond Quran, with some six thousand facets or noble verses, is both the mystical map and the incandescent continent of Divine Mercy whom we call Muhammad. The culminating Messenger of Allah is the living, breathing, eternally subsisting Quranic chapter *ya sin,* known as the heart of the Quran or as the Quran of the perfect human heart. The Prophet's subtle body, breath, speech, awareness, heart, mind, soul, and spirit are crystal clear with the radiant clarity of Quran. The Book of Clarity, the Divine Clarification of all possible worlds in the light of Unity, is the reservoir and safeguard of all wisdom and all beauty revealed through the Prophets that Allah has awakened within the human family throughout time—may perfect peace be upon them all.

The subtle human form, consisting of pure resonance and radiance, is composed from the mystic calligraphy of the Universal Quran. This is the secret Muhammadan nature of every precious human being. All souls were created before eternity from *nur muhammad,* the Muhammadan Light of praise and submission, the column of clear radiance that is the axis and marrow of all worlds and that existed from before all worlds. This First Light is the Divine Luminous Wisdom, which is the Light of Quran, the living Light of every revelation in the entire history of prophecy.

Allah is Mercy, His Messenger is Mercy, His Message is Mercy, the month of His Message is Mercy, and the recipients of His Message are composed essentially of His Mercy.

The secret of Divine Mercy is that It eternally overflows with generosity, constantly conferring radiant gifts beyond conception upon all living beings—on the earthly plane and on the heavenly planes, as well as on the supreme plane of Paradise. This spontaneous and infinite pouring forth of riches is precisely the Fountain of Essence, and its countless gifts are none other than the numberless Divine Energies from which all souls are woven, each soul a unique and eternal tapestry of unimaginable subtlety. From the Divine Attributes alone is creation woven as a mystical tapestry, as a diagram of Unity. There are no random events. There is absolutely no merely material existence apart from Allah the All-Living One. All manifestation is Divine Life, for only Divine Life exists. Thus creation, both temporality and eternity, is the timeless outbreathing of Divine Mercy.

Ramadan is the state of consciousness in which the boundless generosity of Essence displays Itself openly. For an entire month, Ramadan generates a crystal clear reflection of Paradise upon the mirror of the earthly plane. In this mirror of space and time shines precisely the same Light of Allah that floods the seven heavenly planes and that surges within the Divine Heart. Ramadan opens the gates of Paradise. Allah fully manifests Paradise during the Night of Power, and the month of Ramadan is all Night of Power—a single diamond of Divine Peace and Power turning slowly, sunset by sunset, each of its twenty-nine brilliant facets a mystical night, during which unique blessings descend into the entire universe from the Most High Who is the Most Near.

The central core of the month, the precise Night of Power, extends timelessly in earthly time from sunset to dawn. This mystical night, containing the golden midnight sunrise of Essence, actually dissolves and resolves galaxies and subtle heavenly worlds into pure Paradise. Paradise is not a physical, mental, or spiritual location within creation, but the very Heart of the Creator. Thus Ramadan reveals in daily experience—whether to advanced mystics or to beginners in the way of Islam—the inconceivable identity of the primal Essence, the Divine Attributes, the eternal souls, and the pristine creation. This supreme identity, or indivisible unity, is prior to and beyond the mystic union of the soul with its Lord, prior to and beyond Divinity and humanity, Creator and creation—for these dynamic polarities are simply the play of Divine Energies. The Night of Power is the perfect Peace in which all dynamics disappear. O Night of Power, please manifest continually as the very nature of Reality, as the very nature of infinite Generosity, as the very nature of the fully evolved Muhammadan humanity.

We greet each other gratefully within the holy month, within this virgin womb of Divine Luminous Wisdom, by speaking the blessed words *ramadan karim*. These words indicate that Ramadan is the purest manifestation of the particular Divine Energy, *karim,* the Spiritual Nobility that offers Itself generously to all humanity. In desperate longing to express just a single drop of meaning from the radiant ocean of Ramadan—the stormy ocean of love that we feel surging in our hearts during this most mysterious month—where can we turn but to the Glorious Quran? What other than the Book of Clear Light can express the boundless radiance of Ramadan?

The gracefully unfolding Prayers of Ramadan, twenty cycles of prostration repeated for twenty-nine nights, reflect the eternal Quran in time, just as pilgrimage to the Kaaba reflects the boundless Quran in space. The true Kaaba, however, is not a spatial location on the globe but the secret heart of humanity. Just so, the true Ramadan is not a temporal location within the planetary year but the state of fully awakened, evolved, and enlightened consciousness. This is why to practice Ramadan, its fasting and prayer, its acts of kindness and ecstatic delight, gives us a genuine glimpse into the minds and hearts of the gnostic saints who have attained the spiritual station of Ramadan—throughout temporal experience, within eternity, and as the essence of Reality. Ramadan is the experience of the perfected friends of Allah who have lost their personal will in the Divine Will, whose fasting has become feasting.

O Glorious Quran, we supplicate Allah to reveal the nature of Divine Generosity through Your magnificent and mysterious Sura Rahman, so that we can approach and embrace the nature of Ramadan—not through our limited vision or will but through the royal gates of the Holy Quran, which are precisely the Gates of Paradise that open wide during Ramadan. The intense Prayers of Ramadan—twenty cycles of prostration, twenty recitations aloud of Sura Fatiha, which contains the entire power of Quran—lift devoted souls to the level of the great mystic knowers. Their very breath and bloodstream become Quran, these mystics whose only treasure and sustenance is Quran, who are guardians and revealers of the most secret and mysterious lights and meanings surging within the glorious ocean of Quran, who are already resurrected in Quranic bodies with Quranic minds and Quranic hearts, with Quranic senses, Quranic instincts, Quranic movements, and Quranic fragrance.

Ramadan is simply Quran embodied. The transcendent Awareness of Allah awakens through the most sublime conscious instrument, humanity, who is nearer to Allah than nearness and even nearer than that. The Voice of Truth is the Voice of Allah is the Voice of Revelation is the Voice of Quran is the voice of our true humanity—the Ramadan love song of Divine Love returning through Itself to Itself.

Sura Rahman reveals the ever-ascending nature of the soul, which is unveiled during the astonishing month of Ramadan. The Sura of Mercy shows how the soul attains to more and more profound appreciation of Divine Generosity, to greater and greater thirst for Divine Knowledge and hunger for Divine Love. At each new and unexpected level of ecstatic wisdom, the Generosity of Reality becomes more overwhelming, and the soul must become more expanded in order to receive this Divine Generosity. Each new spiritual level is inconceivable and indescribable in terms of the former level, no matter how exalted the former level may have been.

The treasure of Divine Love is infinite. The soul ceaselessly ascends into this Infinity, which it perceives as pure Divine Generosity. Allah draws our concentrated attention toward His Generosity no less that thirty-one times during this short Sura of seventy-eight verses. This is precisely the mood of Ramadan. During the holy month, we cannot draw even one of our forty thousand daily breaths without an overwhelming sense of gratitude to Divine Life, nor can we break our fast at sunset without being blissfully drenched in the downpour of Divine Mercy that refreshes all worlds.

During the month of Ramadan, the universe becomes Paradise, our bodies become Paradise, our breaths become forty thousand Paradises every day, our food and drink become the mystical fruit and wine of Paradise, our love the love expressible only in Paradise, where countless billions of souls radiate continuous mutual affection in simultaneity and total interpenetration. This is the Paradise of Divine Generosity.

During the month of Ramadan, we enter completely into the living Quranic Sura Rahman—our being enveloped, permeated, lost, and rediscovered in Divine Mercy. Gazing everywhere into the essential Energy of Mercy, our gaze, which is the gaze of all humanity, returns to us dazzled by the perfection and justice of Allah and His Creation, our voices repeating spontaneously the sweetest words of wisdom and love, *la ilaha illallah muhammad rasulallah*.

This one true Ramadan will never end. We are forever within its tender embrace, its timelessly unfolding generosity—a single dawn, a single sunset, a single voice intoning a single prayer, a single breath, a single vision, a single Pen, a single Throne, a single Paradise, a single Love, a single Soul, a single Essence, a single priceless Pearl.

O Light of Guidance, You are my life. *amin*

MUNAJAT Composed by Abdullah Ansari almost one thousand years ago in the geographical area we now call Afghanistan, this hymn clearly expresses the most radical insights of Sufism. Yet Abdullah Ansari was famous as a fervent practitioner and protector of the *sharia,* or Sacred Law. Nor would he accept any innovative ideas from rational philosophers, Muslim or non-Muslim. His poem is therefore included in Part One, *Traditional Islamic Resources,* to confirm that the most startling Sufi attitudes are fully consistent with the basic fabric of Islam. These English verses, reflecting only a portion of the more extensive original, were composed to be sung to one of the sanctified melodies cherished by the Jerrahi Dervish Order of Istanbul. Seventy of these precious melodies, their lyrics in English by the present author, and an album of their performance in community is forthcoming from Pir Publications as *Chamber of Mystic Hymns*.

10
Munajat
An Ancient Dervish Hymn

In Your painful mystery of love,
lovers find the cure for pain,
wandering through the desert of longing,
crying aloud, *Allah, Allah.*

Drunk with delight, censored by men,
eyes aflame with desperate love,
Your friends cry out like Moses on the mountain,
"Lord, reveal Yourself to me."

When You consume a heart with Your Love,
You scatter its ashes on the wind.
What need have we for separate identity
when we know You as the only Truth?

Ever present, how can I search?
Always aware, what can I pray?
When You glance with infinite Mercy,
how can any separation be?

We speak only to praise Your Name;
we seek only to express Your Delight.
Ruined are we, annihilated we;
slain by You, we are happy to be slain.

Paradise, brilliant and pure,
without Your Face is worse than hell.

A single glimpse of Your true Essence
brings to naught every heavenly delight.

When there dawns a glimpse of Your Love,
love of finitude fades away.
What drunken hearts You give to Your friends;
looking at themselves, they see only You.

Free us from faces, revealing Your Face;
free us from doors, opening Your Door.
If You but once call me Love's slave,
my bliss will surpass the bliss of Your Throne.

Your perfect Mercy cancels our faults;
rings of slavehood adorn our ears.
You raised us up before time began;
we are Your guests: treat us as You will.

What do I care for the play of Paradise?
You make my every glance Paradise.
When You gaze into my eyes,
no duality can arise.

What this poor one desires of You
is more than the wealth of a thousand kings.
Every life seeks blessings from You;
I am asking for You alone.

This *faqir* has no name and no shame.
This poor one knows neither peace nor war.
True rank is never turban and silken robe:
royalty springs from the heart of light.

Creatures are drunk with Your wine;
I am drunk with the Wine Bearer.
Their drunkenness is the life of praise;
I am lost in Your Silence alone.

You are what they praise at the holy Kaaba;
you are what they seek at temples and shrines.
I am free from every religion,
crying aloud, *Allah, Allah.*

You place a pearl on Adam's brow,
dust of rebellion on Satan's head.
Satan seeks You and finds but himself;
Adam seeks himself and finds only You.

Creation is hidden within Your Heart,
yet it remains unaware.
You are concealed within our heart,
still we remain unaware.

All will is Yours. What can I will?
I want no thing. I want no world.
I cannot care for heavenly delight,
seeking only the delight of You.

What must I do to merge into You?
My heart's blood streams through my eyes.
I hold no key to unlock Your door,
madly crying, *Allah, Allah.*

Whoever knows You encounters Your saints,
cool water for the thirsty soul.
Heavenly states for them are like thorns;
they seek only the rose of You.

Pious acts make the mind proud;
forgetfulness brings repentance.
I rejoice in the night of separation,
for the dawn of mystic union is near.

You exist. I exist not.
Everyone is lost in You.

You are closer than my soul.
This is all that need be known.

This poor one is mad with true joy.
This *faqir* is king of all worlds.
Receiving just an atom of Your Love,
every king and world disappears.

Adoring You has destroyed my ease;
now I rejoice in tests and trials.
Just a mirror for Your Reflection,
I am never separate from Allah.

Love's union is calamity;
its agony is sheer delight.
I neither possess what I used to know,
nor know what it is I now possess.

You are the goal; Your friends are the way.
Earthly fires are drops of dew.
Without seeing You, Paradise is prison;
pleasure is poison without Your Face.

If You bring us to trial, we have no defense.
Grant our lives Your Own Purity.
Give our hearts desire for You only.
Resurrect us through Your Grace.

There are no shores in Your ocean of light.
Your mystery can never be revealed.
Lend us Your speech to thank and praise You;
we have nothing of our own to give.

Heart beats only with affection for You;
lungs breathe only with love for You.
From roses above my tomb
streams the fragrance of fidelity.

For free You create. You sustain us for free.
You are not a merchant of gems.
You never sell the precious Resurrection,
freely granting union to all lives.

Submission is Your luminous way;
our actions are Your Grace alone.
Remove from us the sense of obligation;
never veil us from Your Face.

When I remember You, sorrow is joy.
This is my secret door to Truth.
Without the taste of Your Identity,
praying and learning are illusory.

My direction of prayer is Your Face;
my victory banner reads, *All is He.*
Paradise is not my concern;
hell is not my anxiety.

Beginners on the way speak of the Truth,
but lovers possess neither word nor voice.
Those who desire this world are insane;
those who long for heaven are inane.

Fasting, vigil, pilgrimage
are labors of the wage earner.
Fall in love, the royal crown of flame.
"I am Truth," cries Mansur al-Hallaj.

Live in ecstasy but never speak aloud.
To find Truth, become afflicted by love.
Know what remains beyond annihilation
when there exists neither fasting nor prayer.

My eyes are weeping with love's agony
as I dance on the mystic way,

possessing nothing, not even myself,
crying out, "Poverty is my pride."

This traveler on the true path of love
tastes the peace of poverty,
liberation from personal existence.
What great bliss it is to be free.

The mystic guide opens the way.
Brilliant space surrounds the soul.
The rose of light, its petals Divine Names,
blossoms in the grateful heart.

The mystic guide pours the true wine,
his tears of love flooding the world,
his blessed face, a golden Quran,
shining as the rising sun.

Strive to become the true human being:
one who knows love, one who knows pain.
Be full, be humble, be utterly silent,
be the bowl of wine passed from hand to hand.

PART TWO

Sufi Inspirations

LEAP OF THE DERVISH This poem was composed in Spanish and brought as an offering of love by its author on his first visit to Mexico City, where he had been directed through the dreams of dervishes to establish Sufism on Mexican soil and within the fertile heart of the Spanish speaking world. The absence of much grammatical structure in this poem reflects the fact that its author did not speak a word of the language when he composed it, relying simply on vocabulary lists compiled from a Spanish-English dictionary. The surprising results of this linguistic and spiritual experiment still enchant hearts at the Mesquita Maria de la Luz in Mexico City.

11
Leap of the Dervish
Invitation to the Path of Love

Allah does not place upon any soul a burden beyond its capacity.

—Holy Quran 2.286

Repeat without ceasing the name of Allah and celebrate His praises night and day.

—Holy Quran 3.36

The dervish is a fertile field
cultivated by Divine Power.
Advancing step by step
through seven stations of wisdom
along the way both arduous and joyful,
making every sacrifice,
the dervish completes this steep path,
this royal road.

The leap of the dervish
is the radiant storm of love
filled with exultation.
This heart becomes exposed
to the splendor of Divine Light.
Each breath becomes an offering.

The dervish is a mystic bride
promised to the Supreme Beloved

Who abides deep within all hearts.
This leap is the wonderful wedding
of the soul with Divine Tenderness,
Who surpasses all imagination.
The eternal spouses, lover and Beloved,
now enter the Paradise of Unity.
This is the astonishing matrimony
between our human reality
and the Source of the Universe,
mystically consummated on earth.

The dervish is a holy sacrifice
who trembles in every cell,
in every pore of the skin—
a sacred offering to Divine Love,
melting irresistibly
into pure Presence.

The dervish is also ordinary,
frank, truthful, affectionate—
above all, a person of hope,
a living sign of the eternal love
shared by Divinity and humanity.

This heart always remains
secluded within Allah,
the most beautiful retreat.
To be in continuous prayer
surrounded by essential silence
is the soul's natural way of breathing.

The dervish chooses to abide
neither in convent nor in cell
but in joyful companionship,
friendship with humanity.
Conscious creatures are
invited guests of the Beloved,

called together by love,
drinking from the spring of love
in Paradise and on earth.

The dervish leaps
into the depths of the heart,
plunging deeper and deeper
into the stormy ocean of love.
Apprentice of mystic love
along this arcane way,
the dervish polishes the mirror
and gazes into Essence.
The dervish heart becomes
a blazing fire of love,
spreading light and warmth
throughout creation.

This companion on the path
is a person of passionate intensity
and profound peace,
one whose very existence
has become a sacred vow,
a response to the Divine Call,
an act of total responsibility.

Before this leap, weeping
and repentance are necessary.
Tears of separation and longing
are the ultimate thirst for Reality.
The supplicant prostrates
with grand and sublime desire
until the rain of rapture arrives—
the inundation of Divine Grace
that creates maturity,
mastery, and insight.
At last the dervish leaps into completion,
rises as the full moon of understanding,

and abides in plenitude,
the culmination of the way.

This heart is the universal message
of continuous mutual affection.
In love with Divinity and humanity equally,
the dervish loves and only loves,
reaching the apex of love,
sensing the vertigo of love.
The final dervish vision verifies
true humanity as unanimity.
Human beings are already unified
by Allah within Allah.

The dervish practices peace and clemency,
gifted by the All-Merciful One,
with uprightness and knowledge of unity.
Adept, well versed,
filled with pristine energy,
the dervish heart is a crystal,
the dervish breath a prayer.
Surrounded continuously
by pure souls and angelic beings,
the dervish lover clearly receives
Divine Promises and Warnings.

Friend and helper of Allah,
this eternal disciple of Truth
is accomplished in healing—
interpreter of dreams and omens,
midwife, spiritual guide.

Liberated from every yoke,
this lover penetrates the nucleus.
The dervish mantle is prophecy,
the sole dervish task to be
eyewitness of Divine Light.

Intimate of Allah, consumed by love,
this heart exists only
as a sublime order from Reality.

As the dervish leaps across the threshold,
Paradise becomes an earthly adventure,
not an encounter beyond the tomb.
This leap of the dervish
unifies heaven and earth,
confirming the convergence
of the soul with its Source,
the sunray with the sun,
the reflection with the original.
Allah is the Light of heaven and earth.
There is no other illumination.

This person of profound courtesy,
long-suffering and full of mercy,
possesses the master key of ecstasy,
pouring forth profuse blessings
from the precious lineage
of the most beloved Abraham,
Moses, Jesus, and Muhammad,
pearls of the universe,
clairvoyant and committed,
great waves in the ocean of love.
Precisely like these prophetic ancestors,
the dervish is faithful, trustworthy,
fervent, and blessed.

Both in Paradise and upon earth,
Allah offers the dervish soul
the Divine Welcome,
the garland of love, and the final felicity.
The eyes of this heart are now open.
Nothing is obscure; all is unveiled.
The eternal festival of love

is manifest, perceived, and known
without doubt or illusion.
Here, Divine Sustenance
flows for all creatures.

Made valiant by Divine Power,
this heart opens fully
to the touch of Divine Light.
Passionate and untiring
in aspiration and commitment,
this person of open spirit,
gifted with spiritual senses,
becomes whole and real.
Never depressed or apathetic,
this heart remains united
with its Supreme Beloved.

The secret hiding place of the dervish
is within Allah Most High.
Moment by moment, this lover
chooses Divine Splendor,
eyes flooded with radiance.
Lost in the Beyond,
the dervish spins like a top
that melodiously hums,
You! You! You!

Inebriated, swaying with ecstasy,
submerged in the surge of love,
this lover is transmuted
by Divine Alchemy.
Mundane concerns disappear.
The dervish whirls magnificently
in the Divine Whirlwind.
Such is true spiritual courage,
the leap of the dervish,
this celestial pilgrim

transplanted onto earth,
who remains devoted, reserved,
modest, and harmonious.

Without inquisitiveness or negation,
the dervish celebrates Divine Necessity.
Never rebellious, forever festive,
this lover lives everywhere
by living Divine Life alone—
loving the whole creation,
including every daily event,
by perceiving all as the Mercy of Allah.

The dervish manifests each instant
as the lightning flash of Truth.
Absorbed and transported,
the dervish cries out,
True! True! True!
and therefore truly exists.

This is the timeless pilgrimage
toward Allah within Allah.
Never exotic, the dervish
embraces the ordinary.
Never self-absorbed,
the dervish heart lives for others.
The sole sensation of this lover
is the certainty of the Divine Will.
This leap is human life
perfectly submitted to Allah,
lived freely in the wilderness,
rooted in mystical soil,
flowering in the Divine Heart.
This leap is the spiritual royalty
of every human soul—
the final realization,
the purpose of humanity.

Moving beyond personal will,
escaping from the labyrinth,
there remains no superfluous movement
in the dervish dance offered only to Allah.
This ceaseless contemplative vigil
brings the beautiful Divine Names
alive in every cell of the dervish,
the crucible for future humanity.

Completely in love with Divine Love,
the dervish heart becomes simplified.
Perceiving everyone affectionately,
this heart is integral,
affable, sympathetic,
patient with Divine Patience.
Shining like a dove of light,
this heart is a golden scripture,
a ray from the sun of wisdom
that neither rises nor sets.

Never absent from the Presence of Allah,
never forgetful of Divine Omnipotence,
never distant from humanity,
the dervish is a healing gaze,
a living word from Allah—
sacrificial altar, secret arc,
sacred plowshare,
key to the mysterious code of love.
This guardian at the gate of love
venerates love and burns with love.
Pregnant with love and by love,
wandering ecstatically through love,
this heart resounds eternally
with the music of love.

With intense solidarity,
the dervish prays for humanity.

Spontaneous, liberal,
generous, and daring,
this godparent of all living beings
is their spiritual liberator—
archer of love, ardent fire of love,
adamantine pledge of love.

This lover senses intimately
the bond that unites creatures
with their unique Creator,
racing along the direct way
that joins all hearts.
This lover recognizes instantly
the unbreakable link
between travelers on the path.
Struck by the lightning bolt of love,
executed and obliterated by love,
the dervish lives continuously
in the tremendous earthquake of love.

Allah is in desperate search
of the rare dervish heart.
By this Divine Quest alone
the soul is fused with infinite light
in the mystery of Divine Solitude.

Divinity resplends within this heart
without future or past.
This is the ablution of radiance.
The very foundation of the soul
now melts and flows.
The worker bee is transformed
into the queen bee
in the hive of Original Unity.
The fragrant honey is
the tenderness of the Beloved—
amorous words, mystical caress.

The body of the dervish
now functions with the facility
of pure Divine Love.
Flesh transforms into spirit.
This soul now sees creation
as facets of a whirling diamond.
There is no more dream,
no more dreamer.
Nothing is low or finite.
There is only Divine Revelation
and the rapture of the soul.

The dervish transmits Truth
without egocentric bias,
without any mask,
never mumbling instinctively,
me! me! me!
Poised in equanimity,
this soul is neither apprehensive
nor possessive, perceiving
life and death as synonymous,
perceiving all existence as Allah.
The dervish assists society egolessly,
offering the supreme assistance of love.

This surprising leap of the dervish
occurs in a supernatural moment
like a spiritual somersault.
This mystic leap is an immense merit,
a cascade of luminosity,
a magnificent favor from Allah.

Springing forth directly
from the primordial Source—
firm, strong, humorous, grateful,
free from superstition and supposition—
the dervish is a living synthesis

of every religion of humanity.
This heart is a vine
from the garden of Paradise
that flows with spiritual sap.
This heart is the freshness
of the vast arbors
and radiant shade of Paradise.

The dervish is the interpretation
of the dream of existence.
This lover communes ceaselessly
with ineffable Divine Delight
in the secret palace of solitary Essence.
Always vigorous and active,
this eternal friend of the Supreme Friend
soothes, clarifies, and enlivens all hearts.
The dervish is the authentic healer
of the principle human malady,
the wound of forgetfulness.

The dervish is a shooting star
in the firmament of love.
This heart is the focus of revelation,
the sacred meeting place
between humanity and Divinity.
Never desperate, always serene,
balanced, impartial, just,
the dervish is a clear mirror
reflecting only the Divine Knower.

This divinizing dervish leap
is the miracle of every moment,
every breath, every heartbeat.
There is no longer linear time.
The dervish is the child of the moment
that transcends past, future, and present.
This astonishing instant,

living, luminous, spacious,
is the mysterious temporality
that amorously embraces eternity.
Here in the moment forever ceases
the tyranny of worlds and egos.
Here love alone reigns
over the region of Divine Rest.
Mourning, anguish, and bitterness disappear.

This royal realm is right here,
already englobing creation—
sweet its flavor, exalted its pleasure.
The dervish enters and is absorbed
in the perpetual springtime of love,
without obstructions or obscurations.
This is the advent of Oneness,
the flight into Aloneness.

Benevolent and benign,
fraternal and global,
this heart of obedience and serenity,
love's own work of art,
falls in love with love.
Guiding the ship of love
for beloved sisters and brothers,
the dervish is a dolphin
dancing in the ocean of love,
a red feather twirling
in the tempest of love.

Reality is a prism, generating
in the void a great rainbow
composed of religions and visions,
opinions and propositions.
By permission of the All-Merciful,
the dervish remains within the prism,
within primordial and perpetual light.

Here the Most High teaches ceaselessly
and the dervish listens continuously.

This supreme station,
this step beyond mystic union,
is the banquet of love,
the meeting with Benevolence
free from the danger of self-deception.
This is playful seriousness,
sober drunkenness.
This is the ultimate luxury—
Divine Simplicity.

The heart intoxicated by Divine Beauty
now tastes unsurpassable ecstasy,
interlacing, harmonizing, and attuning
with Allah the Beloved.
Taking refuge in Truth alone,
this heart is the triumph of love,
the treasure and throne of love.

Through this energetic heart
flows the sap of conscious union,
which becomes sweetness
and consolation for all creation.
Body becomes a live coal of love,
bloodstream the ecstasy of Divine Unity,
breaths a string of pearls,
a brilliant rosary of praise.

The plenitude of dervishhood
is simply to be Truth.
This way is a maternal mystery,
the dervish a child of Divine Light
who lives in holy infancy,
a beatitude free from vacillation.
This is the resurrection of love

in a new body composed of love.
Worshiping in the universal temple,
illumined by timeless dawn,
embracing all with unitive knowledge,
the Supreme Beloved alone exists.
Divine Life alone lives
through and as the dervish.

This vertiginous leap
of initiation and transformation
is now complete.
Comprehending deeply,
beautifully equilibrated,
communing with human and Divine,
always tender and communicative,
the dervish attains fullness.
This heart is the flight
of a sacred arrow.
The dervish heart becomes
sanctuary for universal prayer,
guardian of humanity,
aromatic rose, celestial flute,
psalm, staff, balm,
and nourishment for souls.

The dervish lover now gives life freely:
serving, encouraging,
curing, caring, offering,
sharing, running, daring, dancing,
sanctifying, sacrificing, feeling,
adoring, gazing, resting,
and waking early.

The realized dervish soul greets all beings
with the sublime Peace of Allah.
These salutations, utterly sweet,
can resurrect dead hearts.

This lover is the mystic dawn
of the full moon of wonder.

Soul stripped naked of veils,
the mundane masquerade is over.
Beginning and end have vanished.
Allah reveals overwhelming Light,
and the soul turns transparent.

Allah ordains every event
simply by proclaiming *Be!*
and it comes to be.
The dervish becomes this Divine Word,
this mysterious thunder:
Be! Be! Be!

CIRCLE OF ENCOUNTER This poem, originally composed in the beautiful Castilian tongue, is a tribute to the mystical circle dancers of Mexico, the Concheros, the People of the Conch Shell, and to the circling and whirling dervishes all over the globe who represent Islamic tradition. Many of the first persons to receive initiation into the Jerrahi Order in Mexico City were Concheros. Some of them learned to take their first steps as infants within the swirling circle of dancers, to the deep rhythm of drum, leggings of walnut shells, small guitars and mandolins made from armadillo shells, and mystic hymns in which Franciscan Christianity and Pre-hispanic Wisdom are subtly fused. The Conchero Mesa of the Niño de Atocha tenderly welcomed the Turkish dervish tradition, which came to them through their sisters and brothers from the North. *Circle of Encounter* is a hymn of gratitude and praise for this union of two mystical Orders from opposite sides of the globe, this universal romance among the lovers of Truth in all cultures who join hands and move in circles of adoration, esoteric wisdom, and mystic union. Certain important principles and secrets of the Dervish Circle are here set forth.

12

Circle of Encounter

Dervish Ceremony of Divine Remembrance

Night and day, consecrate your being to the Divine Mysteries, meditating constantly upon Divinity—invoking, crying out, praying ceaselessly. This will permit you to enjoy the cosmic dream.

—*Codex of Wisdom from the Valley of Mexico*

What is Paradise? The circles of Divine Remembrance.

—*Oral Tradition of the Prophet Muhammad*

Humble lovers of Divine Love
enter the sacred circle
with upright mind and spirit purified.
This circle of encounter,
whirling eternally,
is a prodigious, storm-tossed sea,
a profusion of blessings in present time
from the mystic saints of all traditions—
powerful and living blessings
that emanate from the single Source,
the only One.

These dervish dancers are
initiated sons and daughters
of the mysterious circle.

As they advance together along the profound way,
their progress is without pause, without end.
In this circle of reconciliation,
only love for the neighbor prevails.
Each person manifests truly
and is recognized by the circle
as a unique word of Truth:
not by bread alone does humankind subsist
but by every word that comes from God.

This circle of holy recollection
celebrates a spiritual wedding
among hearts set sail together
upon the ocean of Divine Light.
During this ecstatic voyage,
courageous dervish dancers
abandon the narrow ship of self
and plunge into open radiance.

Pictures of the surface world disappear,
and pristine creation is revealed
as universal harmony, full of correspondences—
a secret garden, flowering and flourishing,
a manifestation of pure Presence.

Initiates in this circle of union
dare to cry with abandon,
"Receive us in Your arms!"
and immediately they experience
the sacred arrow wound of love.

Courageous dervish dancers whirl
around the central pole of existence,
learning to dispense with themselves
and love others completely,
during both sacred ceremony
and daily experience.

The circle is a path of every heartbeat,
always new and astonishing,
a path that creates beauty and tenderness,
that achieves the incessant
renewal of the universe.

The movement of this nuptial circle
penetrates past physical and emotional states,
and the dervish dancers enter
the springtime bridal chamber
of human soul and Divine Beloved.
To transcend every obscurity
is the proof offered by this circle,
already immersed in Divine Intimacy,
where hearts discover hidden treasure.

Those integrated in the circle of nearness
are universal dervishes from the entire globe
invited by the living Truth
to this mystic nighttime gathering.
The sensibility of the dervish dancer
glides through the circle like a swan,
free from limited actions and attitudes.

The Presence within this circle
activates evolution and unification.
The Divine Pressure of this circle,
rather than any philosophy,
is what creates a diamond
from the inborn nature of humanity.
Come to spiritual warfare,
O warriors who bear the beautiful light!

Hands firmly linked,
voices intone with intensity
the beautiful Divine Names,
forming the primordial chorus.

May we set our hearts on fire,
filling them with radiance and power!

Divine Wisdom, like a seed
sown in the breast of each dancer,
generates through mystic utterance
a conflagration, the sun of the Divine Gaze,
the single gaze shared by soul and Lord,
the pristine awareness *la ilaha illallah.*

The countenance of the dervish
is simple, warm, smiling,
even when lost in adoration,
singing to Wisdom with the entire being,
O Star of the East, grant us your light!

These illumined ones of Allah,
blissful prisoners of love,
move with sheer delight,
borne up on wings of resonance—
the resounding of Divine Attributes
through the sweet sound
of human breath and voice.

This circle of encounter
gives birth to children of light,
the elevation of the dervish dancers
to a height that has no measure.

This circle of remembering
simply calls *Allah Allah Allah,*
and Reality blossoms within these souls
who throw themselves into calling,
becoming lost in this circle,
this chalice completely filled.

This perpetual circle displays

creation returning into Creator.
This dynamic circling is the conquest
of body, mind, and heart,
the victory over separation.
This circle of Allah Most Near
speaks the language of Divine Love,
which alone awakens the human spirit.
These dervishes are leavening
for the bread of humanity.
Transmuted by the alchemy of the circle,
their bodies become stars on earth
reborn with every breath,
their souls a cascade of gratitude.

The practitioners of this global dance,
disciples trained in the liberty of love,
are forerunners of true humanity.
The green branches of their mystic Orders
bear precious golden fruit,
the vast inheritance of humankind.
The community of universal dance
contains the essential code of Truth.
This circle of encounter
is the rootedness of humanity,
this harmonious resonance
the Divine Conquest of humanity.

The inheritors of this ancient circle
abide at the refreshing confluence
of every sacred lineage,
surrounded by every banner,
chanting, *Come, let us follow the Way!*—
the patient steps of our ancestors
along the path of hope and confidence,
the long journey of the human spirit,
its liberation from anguish, narrowness,
and the egocentric battle for survival.

This circle is the transformation
of our clay into mystical fire.

Now the dervish dancer,
who belongs only to Divine Love,
encounters the uncreated essence of the soul,
tasting the ease of Paradise.
Casting overboard the ballast
of all worlds, all self-love,
the circle becomes open space
without any horizons,
the goal of the spiritual journey—
establishment in Allah.

This life-giving circle is no game
but the dimension of sheer aliveness,
the consciousness that responds to Allah,
obeys Allah, salutes Allah—
the vocation of loving the Beloved.
These souls awake in Allah
drinking deeply from the cry
Allah Allah Allah
that comes forth from the amazement
surging in the circle,
this illumined heart
of the body of resurrection.

O circle that shapes the universe.
O circle where the universe submits.
O circle of those with fervor and intensity
who awaken fully within daily existence.
O circle of Messengers, soothing balm.
O circle of Divine Majesty.
O circle pregnant with spiritual gifts.
O circle whirling beyond concepts.

O abode of those who empty their hearts

of heavens and worlds.
O circle of the awakened gaze,
focused solely upon Reality.
O circle of Divine Generosity,
Divine Spontaneity.
O circle of the luminous wake
streaming from souls returning to their Source.
O dervish dancers—
excellent, excelling, exceptional.
O planetary wheel of the lovers of Essence.
O grand commotion within the universal heart.
O wonderful circle of sheepskins
on a polished wooden floor.
O circle of knowing feet.
O circle of life,
whirling since the beginning.
O conscious constellation of atoms
empowered by *Allah Allah Allah.*

O perfect reciprocity between beings
and the Source of Being.
O unexpected gift.
O blessed sleeplessness.
O lightning flash of the leap
to new spiritual levels.
O sacred vestment of light.
O expansion of the heart
into sublime absorption.
O direct and open way.
O waterfall of drunkenness.
O primal innocence.
O permanent place for prayer
within the nucleus of awareness.
O circle of celestial swiftness.
O circle of Divine Breath.
O famous tavern of mystical dreams.
O grateful captives of Supreme Love,

humble servants of the Supreme Cause.
O circle of the representatives of Allah.
O circle of epistles
written by Allah to Allah.

O circle of encounter,
gather us in your embrace!
Intensify us, test us,
absorb us!

COUNTENANCE AND HEART OF THE SHAYKH
Composed spontaneously from long lists of rich Spanish words, this unexpected epic poem manifested as a portrait of my beloved Shaykh, Muzaffer Ashqi al-Jerrahi, drawn from unknown depths of my heart through the process of writing in a language unfamiliar to the surface mind. Both his *countenance,* his beautiful outward form and gracious mode of life, and his *heart,* the invisible secret recesses of his realization, become accessible through these verses. No photograph or film could achieve this fullness, which can only come through words sent by Allah Most High. This poem contains many unspeakable secrets of passionate mystical love, unfolded by Jelaluddin Rumi in his longing for his beloved Shems, the perpetually rising Sun of Love.

The original Shaykh of Islam is, of course, the noble Prophet Muhammad, may peace be upon him. Communities of true lovers ever since have gathered around the shaykhs, or mature mystic guides, who are rightly called *inheritors of the wealth of the Prophet.* The shaykh is not a personality but a spiritual principle—one in essence, dazzlingly diverse in manifestation.

13
Countenance and Heart of the Shaykh
Portrait of a Sufi Master

The heart assumes the qualities of the Beloved.
—*Rumi*

In the bosom of the sacred edifice of humanity
abides the secret of spiritual architecture,
the keystone of love, the Shaykh,
noble soul, sovereign of mystic union.
Always longing for Allah Most High,
eternally offering himself to the One Reality,
the Shaykh is surrounded by dervishes captivated by love,
his intimate allies in the closest friendship
and most elevated ideals.

Above the noble pillars of religion,
beyond formulas of faith, legends, and allegories,
shines the Shaykh, example and guide for souls.
Pinnacle of knowledge and refinement,
subtle model for the mystic way,
he essentializes human activity,
spiritualizing all phenomena
and making the human heart fragrant
with the litanies of his breath.
The Shaykh is full moon of Truth
and living word of Truth.
Ardent lover of Divine Peace,

adorned by spiritual humility—the mystic poverty
that possesses nothing and claims nothing—
the Shaykh prunes the tree of personal will
and tends the nursery of souls,
experiencing great love
for future generations.

Heart reposing in Allah
and feet rooted upon the planet,
the Shaykh is an oasis of the One Reality,
a transmutation of the ordinary world.
Living book of mystic dreams,
the Shaykh is the diagram of the Divine Plan.
He is a rich vein of gold.
He is a Golden Age.

Fully realized, dressed in common cloth,
free from the weight of personal will,
this enamored one of Allah and leader of humanity
soars in ecstasy without the slightest effort,
borne up by the Divine Wind alone.
Divine Light arises in his heart,
overflowing from his entire body.
Free from the nightmare of the self-centered world,
relieved of self-love by the inscrutable Will of Allah,
the Shaykh falls in love with all conscious beings.
By sharing their extensive suffering,
he enters the presence of the Beloved.

The Shaykh is the polestar of Love,
portent that shines far above
the narrow streets of the limited world.
He is the human soul released into freedom,
showing forth the image of Divinity
that is hidden in humanity,
that sacred image never tarnished.
The Shaykh, clairvoyant seer,

uncovers the unsuspected potential of every heart.
This excellent emissary of Allah,
always adoring Allah,
bears news of eternal joy for humankind:
the recovery of its conscious unity with the One.
Clothed in the white linen of Purity,
he sits in equanimity upon the stone seat of Truth
beside the spring of Love and the lily of Beauty.
Accessible to those who go in search of the Real,
at morning light, noon, and midnight
he teaches sweet prayers and lucid praises
to wise practitioners and beginners on the way,
to all those with strong predilection
for contemplative practice.
Stirring amazement and gratitude in their hearts,
the Shaykh attracts his loyal friends
to the life of perfection—
a life more alive than the life of the entire universe.

When a physician administers
the precise drop of medicine,
fever disappears; the sick recover and become strong.
When the Shaykh presents the nectar of his teaching,
which cures the loss of memory,
melancholy minds become skylarks—
emancipated,
flying and singing in the sky of wisdom.
The dichotomy between heaven and earth
has now disappeared.
These liberated souls abide with the Shaykh
in the fusion of nearness and distance
without veils of darkness or veils of light.
Free from ambiguity and obscurity,
their life is a holy sacrament.

His heart always falling into ecstasy,
the Shaykh's flourishing existence

is universal call to prayer,
chamber of prayer, light of prayer,
and lyrical cadence of prayer.
Eyewitness to Divine Light, the Shaykh
demonstrates to humanity with great intensity
the reality of Divine Unity,
beyond myth, science, and theosophy.
He redeems human experience
with his spiritual sensibility.
He is like an immense choir glorifying God,
free from apparent forms, from the apparent ego.

Taken captive since childhood by Allah,
tested and trained by Allah,
the Shaykh has made the long journey
to the unthinkable goal of union.
Convinced totally by Allah,
Who is the fundamental Demonstration
shining in every recess of the universe,
the Shaykh receives intimations of Divine Power
flashing in his heart, beyond names and letters.

The prowess of this morning star of humanity
is his ceaseless proclamation of Unity—
not simply through sacred hymns
but through every motion of his body,
through every thought, every intention.
His very existence is to ponder and profess
primordial Reality in the present moment,
a promising spiritual procedure
that has given birth to mystic saints
in all the epochs of human history.
The Shaykh is the rich legacy of Divine Love,
the ancient caravan bearing indescribable wealth:
singing birds, caramels, love letters from the Beloved,
and all the other exquisite gifts of Allah
to the secret heart of the human being.

Self-love manipulates and dominates,
propelling the conventional world toward blindness.
But when touched by the Shaykh, lightning of Allah,
this obscuration, this worship of the ego,
melts into fathomless ecstasy.
The spiritual leader who with this valor and capacity
transmits the ancient lineage of wisdom
is the true Shaykh, the wonderful counselor,
the beautiful prototype, the messenger of Love.
Mature, wise, kind, open—
he is the impartial benefactor of humanity.
As the eternal discourse of the Divine Word,
this guide becomes head and heart
of the mystical body of the universe.

Actions are the confirmation of the Shaykh.
His liberating and renewing vision
is totally infused by Divine Love,
by transports of spiritual inebriation.
This supreme lover floats lightly in Divine Presence,
immersed in the pulsation of the Divine Heart,
which renders everything nothing
with the blinding brilliance of its rays.
Here is primal door and final harbor,
both way and goal, cocoon and butterfly,
supreme enterprise, sublime discovery,
the spiral evolution of the soul.

Purified by sweet pangs of Divine Love,
clothed by Allah Most High
in magnificence and integrity,
the Shaykh listens only
to the voice of the Beloved,
both lullaby and eloquence,
experiencing mysterious quietude
in this enchanting Divine Presence.
Because the courageous guide

roots totally in Allah, the root of Truth,
his heart becomes a cluster of grapes
plucked by the Beloved in a twilight
permeated by praise,
pervaded by springtime fragrance,
punctuated by rushing streams of Love.

After mystic fusion with his Beloved,
this vagabond of Love, friend of all souls,
wanders through the grand prayer hall of creation,
the sacred temple of Divine Manifestation,
taking delight solely in the invocation of Allah.
The reign of the Shaykh is timeless,
removed from the fleeting quality of life.
The eyes of his intuition have been clarified
by perceiving all religions as spokes
within one wheel of revelation.
This guide abides at the focal point
of the universal circle,
at the still, calm, silent center.
His awareness, empty of vain concern,
free from the consciousness of being a subject,
is an overflowing vessel of Love,
a spring of the Water of Life
forever flowing.

The Shaykh, without desire for recompense,
is praised even by Allah.
He passes beyond the limits of creation
and miraculously attains Divine Life,
above worlds of change, above worldliness.
Such a supreme aspirant
becomes an ambassador of Reality,
an heir to the imperishable wealth of all Prophets.
Each Shaykh is a messenger from Allah
bearing unique responsibility.
Each Shaykh is a sunflower

oriented toward the single sun,
a pitcher of blessedness to fill the cup of humanity,
a kite dancing gracefully in the four winds.
Each Shaykh incarnates an incandescent destiny.

Open entirely to the Divine Decree,
the Shaykh ceaselessly recalls and realizes
the Divine Attributes in the depth of his being,
savoring the tender Divine Welcome
in the supreme Paradise of his heart.
Realizing that knower and known are one,
he concentrates in the most esoteric meditation—
the ceaseless reflection upon creation
in all its vastness and intricacy
as the clear reflection of the Creator,
refreshing and radiant, protean and diverse,
essentially lacking nothing.
This adept leader and contemplative drinks,
here and now, from the unequaled fountain
in that Paradise beyond the heavens,
that Garden of Essence
without hierarchy or redundance.
The immanence of Allah as Creation
is the silent music,
the breeze of the present moment.
Transported by Love's enchantment,
eternal guest of the Supreme Beloved,
the Shaykh, sings melodiously
in the untranslatable language of mystery,
"I know the joy of reclining on the breast of the Beloved,
caressed by his shining curls, perfumed with musk,
the joy of sharing the mystical supper
with the mystical traveler."

Voyager beyond abstinence and satisfaction,
his only itinerary is the path of universality.
Follower of the ancient doctrine of Unity,

the Shaykh rejoices in the One,
breathing the One, merging with the One.
His breaths are prayer beads.
The sacred plow of contemplative practice
cuts through the rich field of his heart,
turning over the soil of the mind into the sunlight.

Divine Force plays through his entire body,
which burns like aromatic incense
lit by Divine Light, the Giver of Life.
His mode of being, as living torch of the tradition
transmittable to successive generations,
revalues the entire limited world, dissolving
the limited perspective, the deception of isolation.
The arid life of men, always threatened with ruin,
miraculously becomes green again
beneath the gaze of this intuitive gardener.
The stream of his daily prayers purifies the planet.
The communal body of humanity
shines with his ablutions of light.
His spiritual attainments empower the novices.
The entire universe with its countless worlds
whirling in great homesickness, aspiring only to Truth,
reverently enters the sacred rite celebrated by the Shaykh,
the nocturnal circle of mystic return,
the circle of Divine Urgency
like a midnight wind.

Profoundly moved by this grand undertaking,
protecting angels surround the dervish lover,
who asks the Shaykh's permission
to become intimate with his circle.
This unforgettable immersion
of the intellect in angelic jubilation
is pure unitive gnosis:
there exists nothing that is not God.
Like waves in a shoreless ocean,

the dervish dancers now
converge upon the Shaykh,
central pole of prophetic Light,
indispensable sign of Truth,
unimaginable abundance of Love.

A global sea of splendor surrounds the Shaykh,
roaring with powerful sonority,
all-present and all-penetrating,
Allah Allah Allah.
Green is the shining water of this circle,
green the robe of the Shaykh, silken and flowing,
green within green, free from separation.
The noble captain and navigator
stands like an eagle in the center of the circle,
indicating the direction for the mystic crossing.
His faithful mariners
experience the rupture of the habitual world,
the blessed shipwreck of ego.
Souls know how to swim in Divine Light.

The Shaykh, submitted servant of the Beloved,
never repeats the past but rises fresh from the Source,
demonstrating with simplicity of heart
the principle of mystic priesthood.
This humble sovereign, this ecstatic mendicant,
married to all lovers of the Supreme Beloved,
sacrifices his personal being and is consumed
by omnipotent fire, the Lord of Nearness,
yet remains unscathed—
a minaret of ruby among inextinguishable flames
in the alchemical furnace of Love.

The master of the circle is the refuge of the dervishes,
deepest desire and salvation of the dervishes.
The Shaykh is the health and holiness of all humanity.
Since ancient times, the sweet salutation of the Shaykh,

as-salam alaykum, may peace be upon you,
expressing his full participation in Divine Peace,
has cured disciples of spiritual forgetfulness,
welcoming them to his mysterious abode,
the sacred totality of Being.

His kiss of welcome is supremely desirable,
his right hand the instrument of Divine Power.
His breath, his palms, the soles of his feet
are naturally perfumed like sandalwood.
The fragrance of his perspiration
saturates the circle of dancers with longing.
His blood sings *la ilaha illallah,* the One alone exists,
and his heart plunges into the ocean of annihilation.
His entire body becomes the honey of Love.
The mist of doubt and guilt dissipates
in the light of his indivisible plenitude.
His gentle, melodious words sweeten the universe.
The Shaykh is a nightingale in the rose garden of creation,
its feathers shining in the rays of the midnight sunrise.
He is a jewel of the mystic East,
a golden book of esoteric paintings,
a lute resounding with imperishable music.

The Shaykh, profoundly religious
without the religiosity of external form,
has been seasoned in the sunlight
of the most beloved Muhammad,
seal and summit of Prophecy.
He distributes the single Light granted
to all Prophets sent to all nations
during all ages of this ancient planet.
The Shaykh is the eye through which Allah
contemplates His marvelous works.
The precious discourse of the guide,
presented in simple words or cultivated speech,
bears the master key, the secret cipher of Unity,

the living axiom of Unity, which confers wisdom
upon the countenance and heart of humanity.

Based only upon unparalleled Reality,
the Shaykh becomes truly serene
without a trace of vacillation,
enjoying only the supreme Meaning,
the Ground of Being, the Uniquely True.
The spiritual aridity of the conventional world,
this place of frightful conflicts and crowns of thorns,
this prison consisting of the illusory notion of self,
comes to an end here and now in the rainy season,
the happiness that does not proceed
from any limited category of knowledge.
Leading and guiding humanity
from his station within the dervish circle,
benign hands open in perpetual prayer,
the Shaykh is lifted into the splendor of intercession.

The sure indication of a true Shaykh,
who divinizes the life of his times
with the power of his heart,
is his similarity to the Prophets—
rope-soled sandals of strength,
staff of daring, sash of freedom,
banner of benevolence, tambourine of praise,
sword that cleanly cuts the knot of ignorance,
shield of equality and impartiality.
The mystic guide is known
by his total submission to Allah Most Near,
conscious submersion in the ocean of Light,
ascent toward Allah, and restfulness in Allah.

Nomad who wanders the expanse of ecstasy,
his beatific vision never obscured
by limited conceptuality,
the Shaykh takes up the poetic word

to inspire members of his mystic Order.
Softly he sings a mysterious elegy,
prologue to the final mystery:
"I hear the sigh and whisper of the Beloved;
unloosed hair streams across my pillow.
I hear the skylark mad with Love
and know the unspeakable subtlety of union
on the wedding night of Divine Delight."

The Shaykh is an ancient tabernacle
in the desert of Divine Access,
place of covenant between humanity and Divinity,
focus of the spiritual diligence of the people
and the spontaneous revelation from Allah,
place of extraordinary inward labors and tasks.
He is a tank of fresh water for thirsty pilgrims,
a sacred pyramid for wise beings of all cultures
who wish to be immersed in contemplation.
In the transport of wisdom, the Shaykh becomes
fully absorbed, performing amazing spiritual feats.
Dancing with abandon in selfless anonymity,
he sings, "Creation in its totality
is the ecstatic loom of One Consciousness.
All living beings are variegated cloth
woven by a single Weaver.
The heart's pulse, reverent affirmation of Unity,
is the life-giving rhythm of a single Drummer.
Systems, titles, customs, and theories are spider webs
trembling in the tempest of a single Love."
Inebriating the universe,
this hymn of praise, this mystical audition,
is the marrow of every cosmogonic myth.

With radiant complexion and clear gaze,
the androgynous Shaykh dialogues with his own heart.
His masculine aspect is generative;
his feminine aspect conceives and brings to light.

Chosen by Allah from before conception,
immersed in Allah's Mercy without limits,
the Shaykh displays great tenacity,
always putting into practice, always experiencing directly,
always moving with subtle firmness and concentration.
His even breath is the rudder,
his focused attention the helmsman.
When his boat of nerves and senses
encounters the Divine Flood of ecstasy,
the Shaykh, forgetting everything that is not Reality,
embraces within his purified, strong-nerved body
the iridescent torrent of Divine Energy.

This person delegated by Allah, skillful and knowing,
shivering and trembling subtly with Divine Love,
whirls at the very center of creation,
waves of *Allah Allah Allah* ringing in his ears.
This quaking of his whole being unveils
the marvel of the stormy ocean of Love.
His heart and countenance shining with Divine Light,
the Shaykh is the renaissance of human grandeur.

This complete communion and union
between human being and Divine Energy
requires a sublime transformation,
accomplished by the Great Transformer.
At this key moment, at this crossroads,
the Shaykh is transfigured into clear Light,
into pure Presence without features.
This is the step beyond precepts
into Clarity and Wakefulness alone,
the final point of the soul's trajectory.
This is the divine threshing of human wheat,
the Day of Judgment with new meaning:
tares are burned in the fire of Love.
Now divinized, the Shaykh becomes translucent.
He is a mystical night bird,

trilling during the timeless Night of Power.
He has become perfumed clover of Paradise,
murmur of wind through fragrant trees of Paradise.
There remains nothing about him separate from Allah.

This height of consciousness, this summit of ecstasy,
this appearing of the Creator
through a transparent creation,
is the fulfillment of contemplative practice,
the goal of the path, the perfection of communion.
The trumpet at the End of Time now thunders in triumph.
It is the sweet cataclysm of Love,
the wise madness of Love.

Dressed as a humble pilgrim,
the conquering Shaykh circumambulates the Kaaba,
the kernel of universal religion.
He is crowned as one who triumphs
over the narrow, anguished tomb.
He is the trunk of the Tree of Life,
the cosmic affirmation *la ilaha illallah*—
there is no reality other than Supreme Reality.
Now he manifests as a benevolent troubadour
traversing the earth, an anchorite in worldly clothing
who sings hymns about the irreality of self,
who holds high the yellow tulip of the Sufi Way.
The Shaykh is the sacred promise of Allah,
never weakened, never broken.
The Shaykh is the restoration of humanity
and the universal resurrection.

In this rarefied sphere of commitment,
awake all night in exceptional declaration of affection
and contemplating the esoteric diagram of the heart,
the Shaykh can kneel on the sheepskin dyed dark blue,
the ancient blue throne of mystery,
seat of power, place of Divine Rootedness.

Here he enters the pantheon of spiritual heroes.
Here in the sweet shade of Paradise
mysteriously cast in the desert of this world,
the Shaykh can wear the green turban of unification,
the green tunic of transformation,
the black belt of authority with its emerald buckle,
insignia of benign power and universal justice.
Smokeless torch, vigorous and bright,
free from curiosity and anxiety,
expressing Truth with surprising boldness,
the authentic sage founds his own Order.
He bears the quiver of Love's sacred arrows.
His eyes are the bow.

The Shaykh abides in the valley of revelation—
not a mythological place but a panoramic vision.
Here he perceives the universe
in its sheer insubstantiality
composed only of Divine Light,
with no illusory dream of a material world
separate from the One Reality,
no false supposition of any second reality.
Here there is no mask, only the Divine Countenance
transcending comparison and description.
Here, with great bliss and opulent lyricism,
the Shaykh celebrates with every gesture of his being
the indescribable diversity of Divine Creation,
timelessly established in the tranquil depths
of Divine Unity, the Majestic Will—
eternally whole, eternally unchanging.
Consciously rooted in this Divine Foundation,
the Shaykh becomes the instrument of Divine Urgency,
the balance of Divine Justice,
the vessel of Divine Tenderness,
the tuning fork of Divine Harmony.
Through the Shaykh, all creations of the One
with love and longing sing in unison,

alhamdulillah—universal praise flows only to Allah.
Sovereign wine and incomparable chalice,
the Shaykh chants the cardinal theorem of Sufism:
"My heart has been made capable of adopting all forms."

Keeping vigil continuously at the edge of eternity,
the Shaykh is a celestial sailing vessel that glides
in the gale of temporality, set sail for the sake of souls.
Clarifying the entire universe, confirming Truth,
he is a stained glass window and a clear crystal,
both Creative Energy and Divine Peace.
Ever-flowing spring of beautiful qualities,
vial that releases the secret perfume of Sufism,
the Shaykh is seer, visionary,
knower of the tablets of universal Torah.
He is the victorious one, never depressed or dispirited,
always dissipating mistaken ways of perceiving,
uncovering instead the equality and divinity of all.

His heart is the virginal womb
where Divine Mystery originates anew,
his presence the delightful greenness of Paradise,
vineyard of the delight and delirium of Love.
Unexpected visitor to the melted heart of the lover,
mystical friend and guide of the souls
who abide in brilliance beneath the prophetic cloak,
the Shaykh is the road of zeal pointed out by the Prophets.
He is the courtship between Allah and the beloved soul.
He is the unsealed volcano of Divine Love.

Never willful, never isolated,
dwelling rather in the cosmic abode of Divine Will,
the Shaykh takes the direction of Truth
without the slightest uncertainty or hesitation.
He is a pure vow, a clear sapphire,
a sweet shepherd's flute.
He strives astutely to awaken the human heart

and even our inert clay, which can become diaphanous
through the influx of the Holy Spirit.
Elevated by Allah Most Merciful,
the stature of this mystic guide is beyond measure—
priceless pearl, crest jewel on the brow of humanity.

"My beloved friends, my allies."
This is the tender word, open invitation
to the banquet of affection,
to the joyful uproar of the lovers,
spoken by this amicable master of Love,
this kind master of good counsel.
His fullness of heart and expanse of spirit
transcend all horizons, all cultural patterns.
There is no analogy for his tolerance,
delicacy, dignity, and mildness.
Totally engaged in the education of humanity,
with his very gestures he demonstrates
the way to encounter Divinity directly.
The Shaykh wears the earring of a slave of Love
and also the royal ring of Divine Command.
This ennobled guide is anchor of humanity,
guardian angel of humanity, touchstone of humanity.
He is our disappearance into universal Being.

Even on this highest plane of wisdom,
the longing of the Shaykh remains excruciating.
It is the unspoken scandal of Love.
His arduous development is never completed.
Desiring only the sweet Divine Consent,
he is crushed by the anguish of Love.
Purified by the blaze of Love,
protected by purity from the venom of the world,
the Shaykh at every moment is expecting
Annunciation, Nativity, and Assumption
within the secret cosmos of his own body,
which is the mysterious obverse of the universe.

In his body the future life of the soul is already present,
and both antiquity and modernity disappear.
This lover of the Real awaits, without respite,
the imminent appearance of his Beloved.
With intense ardor, he loves
to convert his body into Divine Flame,
his ego becoming ashes on the wind.
This is the tumult, the holocaust,
the universal sacrifice of Love.

May Allah Most Tender pity this noble soul,
crowned with flame at the apex of Love,
seated upon the mysterious jade peak
above the seventh level of wisdom,
above Archangels, above all degrees of sanctity.
May Allah protect with great carefulness
this empowered one of Love,
who sings intrepidly to the celestial court
a strange hymn of praise,
syllables planted in the heart of humanity,
a holy document of ecstasy
to constitute the new heaven and the new earth:

"No longer will I plow the sandy soil of personal will
or labor on the estate of profane power.
I neither come nor go on the road of time.
I neither sleep nor die.
I am a warrior of Love, participating
in the final defeat of the forces of negation.
The earring and the bracelet of the Beloved
are my sole weapons, my armor that
neither forgetfulness nor death can penetrate.
My soul has been stolen away
by the aroma of the Beloved,
the sound of his harp, the touch of his shoulder,
the sermon of his breath, the fluency of his tongue.
His mansion of coral is my own subtle body.

Facing directly into Supreme Reality,
I hear and understand Its mystic idiom:
Dare, O human being, to awaken!
Harmonize your song; intensify your commitment.
Consult your heart and your heart alone.
Expose yourself to loving; seek the protection of Love.
To arrive at truly being,
come past the curtain
waving in front of the Divine Light,
which is your own light."

The human seraph of Divine Love continues:
"You alone, my Beloved, are protection,
support, inner force, and true being.
Always courting my soul, always in sympathy,
you are my rapture, well-being, and joy,
the thrill of my existence, the consecration of my life.
I kneel in your presence, heart in prostration,
enraptured and exalted by your essence.
The straps of your sandals I kiss with delight.
I cannot find words to describe my mystical spouse,
the favorite of the universe, the hidden pearl.
O thief of God-filled hearts,
I submit to your overwhelming embrace,
annihilated by this magnificent Love.
Without home or asylum, without path or trace,
without any escape, any remedy,
I am lost in the desert, mountainous and remote,
the flowering desert of your Love.
You are closer than my central artery.
For you I have been mortally wounded by sweet arrows
in the mystical assembly of the heroic lovers of Love.
It is the radiant death of mortality.
I am amazed by this depth of Love,
which removes reason from the enamored ones.
I am an ant on a small leaf of the immense tree of Love,
a mote of pollen in the endless abyss of Love.

My naked heart is a brilliant ember
before the royal blue throne
of the ardent saints of my lineage.
Always ascending, this infinite aspiration
that no flood can extinguish
confirms the mystery of Assumption.
I am the virgin bride of Divine Wisdom."

O soul of the lover, walking your predestined path,
attend this mature master of the Sufi Way.
Come close to the Shaykh. Abide by his teaching.
Stir up your love for the royal leader.
Strive to enter his heart completely,
to identify yourself with his heart.

The mystic guide is an athlete of paradox—
his existence without ego in Allah Most High
is the somersault of a tightrope walker.
To succeed in describing or recognizing him,
to be able to read his message in your heart,
dispense with mundane criteria, finite premises.
His spiritual attraction, never deceptive,
is the most powerful electricity in the universe—
pure mobility, pure brilliance, pure potency.

O soul of the dervish lover, devoted solely to Love,
the religious genius of the Shaykh,
his literary inspiration, and his cosmic vision
are simply his apparent forms.
To dare seek audience with the Shaykh beyond form
is the responsibility of the committed disciple.
At the right hand of the Shaykh of Essence
where all symbols converge and join,
here in the ceremonial metropolis,
here in the hermetic atrium,
lead is transmuted into pure gold,
pure peace, pure silence.

The essential thread
is the flight of the One to the One.

O soul who takes shelter only in Truth,
entreat the Shaykh for the interpretation
of your luminous dreams—
good auspices, evidences of inner light,
which fly like birds through the sky of wisdom.
Entreat the Shaykh to elucidate
the secret grammar of daily events,
the esoteric vocabulary of every living being,
the plan of the universal spectacle,
and the reality of the sole Spectator.
Entreat the Shaykh to unveil
the valor of the martyrs of Love.

O beloved soul, remain alert and prudent,
never removed from the Divine Moment.
Released into the exuberance of Love,
fast and feast in the mansion of turquoise,
the subtle residence of the Shaykh,
enveloped in sweet clouds of incense
that rise to celebrate the worship of Love.
His private door is never locked
to those who marvel at the Beloved,
those impelled by the urgency of affection,
the illumined ones who, with lyric enthusiasm,
move with every atom toward the Origin.
Enter the sumptuous estate of the Shaykh,
encircled by walls of polished marble.
With steps of a deer,
walk his garden path in silence
beside his lotus pond of peacefulness.
Breathe with elation
the perfume of contemplation
from his flower garden of veracity.
Here are no vipers, no enemies,

no twisted minds, no pride or vainglory.
Liberated from estrangement and complexity,
gaze at the clear sky of the only Consciousness,
Origin of the universe, hidden Glory, immense Beauty.
Draw forth this sweetness from your own heart.
Humanity must succumb to Divine Love.

O soul who with great care in every moment
aligns with Truth, prostrates to Truth,
your precious human body is the sheath of the scimitar,
la ilaha illallah, the cardinal axiom of Unity,
the absolute principle of Clarity,
healing balm for every illness,
resolution of every dissonance.
This sharp blade of *la ilaha illallah* cuts easily
through the conceptual spell of opposites.

Submit to the compassionate and healing surgery
of the Grand Shaykh, Nureddin Jerrahi,
the surgeon who handles this sword of light.
The cry *la ilaha illallah*—
there is no reality apart from Supreme Reality—
alleviates suffering, evaporates futility.
Call out with your whole being to Sultan Nureddin.
Become transparent to Divine Unity,
free from burden, sorrow, compulsion.

Seek shelter in the luminous tent of Nureddin,
emblem of the universal religion of Abraham,
open to travelers from all directions.
Retire to his hermitage of constant remembrance
hidden among the dunes of the phenomenal world,
to his holy table, filled with apples of wisdom,
dates of charismatic affection,
words and flowers of wise poets.
Travel with celestial speed in his carriage of royal blue.
Study in his mystical college, his eternal brotherhood,

constituted by Allah's beautiful commentary
known as Nureddin Jerrahi.
In this esoteric school of every moment,
this house of song composed of every breath,
discernment and passionate longing prevail.
The subjects of study are mystical maps,
living manuscripts of the lives of saints,
full of the elixir of knowledge and submission.
Allied and unified with Truth alone,
collaborate with this generous Shaykh;
treasure and ponder his precious teaching.
Receive now the royal commission of Nureddin,
whose rank is the Light of Universal Religion.
Brilliant comet in the firmament of Islam,
Nureddin is standard-bearer of the Religion of Love,
which is free from frontiers.
He is the culmination of sanctity, the choice of Allah,
the incarnation of the mystery of Supreme Love.

O colleague of ancient lovers,
radiate to modern humanity
this timeless radiance of Prophecy
that emanates from the luminary, Nureddin Jerrahi,
instantaneously penetrating all barriers and veils.
Be capable of bringing illumination into practice
year after year in the wheel of years,
always remembering what humankind needs.
This is your urgent task, your blessed fate,
your alchemy for transmuting the world into Reality.

O soul maturing in Divine Love,
develop a new spiritual disposition,
never negating, constantly affirming.
Have no preoccupation with gain or loss.
Interact harmoniously with companions on the way,
encouraging mutual recognition, mutual fondness.
Be pleased with your compatriots in the mystic Order,

this elevated and powerful environment
of the Great Pir, who is fully realized,
venerated by lovers from all brotherhoods
as the axis of mystic lovers.
He imparts his delicate tone
to the hearts of his loyal dervishes,
their acts coinciding with their words,
appearance united with Reality
in hearts that cease to be incongruent.

The gentle comportment among disciples of Love
is a precise reflection of their comprehension.
Their bearing in every moment—
their tender respect and obedience
before all the creatures of Allah—
is the way to confirm their spiritual attainment,
their competence in mystical union.
Their presence of mind, always pacifying
and delighting saddened hearts,
is the protocol of Love, the priority of Love.
These skillful dervishes, close to the Shaykh,
participate with great devotion and attention
in all the precious lives of creation
as well as in Transcendent Life—
in Divine Diversity and in Divine Unity.
There is no substitute for this true praise
for the Giver of Life, the All-Merciful Creator,
for this engagement with His sublime Creativity.
Apply yourself to the service of living beings
and approach the Lap of the Beloved,
touching His clothes of iridescent silk,
His beautiful earthly manifestation.

O dervish taken vigorously by the Shaykh
in his experienced hands, behave affectionately,
with the very generosity of Supreme Love.
Be the contemporary of every human being,

as the Shaykh educates all humanity
with a great gesture of affection on a global scale,
with thaumaturgic power,
overflowing with reason and wisdom.
Commit yourself to being in harmony with the guide,
who is the marvel of Love, the support of the heart.
Bathe beneath the cataract of his love;
be warmed by the shawl of his love.
His endearments are gifts from the Beloved.
The Shaykh acts with great assurance
as head and founder of a mystic Order,
as leader of the plenary conference of Prophets.
He takes the reins of the congress of saints,
representing all religions of the earth.
Extending from the inner sanctuary of Truth,
his right hand of spiritual certainty
is the transparent Hand of Allah,
the One Who alone exists.

O divinely infused soul, rejoice in your Shaykh.
Persevere in Love; remain awake all night in Love.
When taking refuge in the study of ancient wisdom,
never forget the presence of your Shaykh—
his apostolic mission of Divine Love,
his naturalness in love, his constancy in love,
his tender and courteous affection toward all persons,
toward animals and even insects.
Benefit from his love,
the meticulous mercy of his love.
Contemplate with joy the majesty of his love,
the fundamental happiness of his love,
the mystical maternity of his love,
the solar explosion of his love.

O soul who faces Truth alone,
remain profoundly moved by the Shaykh
and you will be able to comfort humanity,

which is lashed by the gale of temporality.
Become formed by the Shaykh,
who is universal knower and consoler.
Consecrate your entire being to the Shaykh,
a career of every breath, every step, every perception,
illuminated by a supernatural sun without nightfall.
Have no fear or limited concern.
Humanize the people; calm the people.
Admonish humanity with your own life of love.
Warn everyone about the snare of the limited self,
the fatal venom of negating others.
Prepare yourself to contribute
your entire existence to humankind,
taking every risk, walking beside every abyss.

O willing soul, O soul who is truly prepared,
O soul entrusted with Supreme Love,
verify directly with the eyes of your heart
the elevation of the Shaykh, his astonishing degree,
and contemplate the consummation of your own humanity.
Enter silent dialogue with the Shaykh.
From this bond and communion
a mysterious blending will occur between your being
and the being of the Shaykh, exalted by Allah Most High.
This is the proven way of the Sufis:
to contemplate the perfection of humanity
in the brilliant mirror of the Shaykh
through the clear light of ecstatic love,
the sole moving force of the universe.
Concentrate your mental faculties at the heart
and fly like a bird released from its cage.

O soul in supplication, O consort of the Beloved,
work with calmness and be content.
The continuity of transmission is secured by the Shaykh,
this wonder working high priest of Divine Love.
The cup of the present is full to the brim

with the exquisite wine of Love,
breast filled with the living gold of Love,
mind with the harmony of Love.
This is the prophesied harvest—
not mere aesthetic vision but the cultivation
of nourishing and medicinal plants
for the well-being of civilization.
The contemporary is no pallid copy of the past.
The point of departure of the Prophets is present.
Copious insights of the heart are surging
in the powerful circle of pure Presence
formed by the lovers of Supreme Love
by chanting *Allah Allah Allah*.

All the esoteric traditions of the earth,
consequence of slow and long elaboration,
converge in the sacred oak grove of the Shaykh.
Hostility is now the only heresy.
The white steed of prophetic energy
is prancing through communal consciousness,
the Shaykh its noble rider.
The treasury of prophetic wealth
is open to the mystic lovers,
those in every culture who possess true hearts.
Sincere neighbors on the planet, full of wonder,
discuss, collaborate, and prostrate before the One,
experiencing sympathetic unity and agreement.

The Shaykh is humble message bearer of Allah,
stately representative of Divinity
and creative current of Reality.
Magnet of Love, mysterious crescent
that grows eternally, the Shaykh
fills the universe with illumination.
Resonating with Divine Names,
the breath of the Shaykh is the cutting edge,
the compassionate weapon of Divine Light

that severs the links of negativity.
Dispensing the sweet nectar
of courtship between lover and Beloved,
this truthful guide uncovers and awakens
the secret chrysalis of every heart,
spherical chamber of rose light
without the least tint of darkness.
Longing to return to its motherland,
the butterfly that emerges from this cocoon,
unexpected by the world,
is the imperishable soul,
its iridescent wings the words of Truth.

To submit as the sacrificial lamb of Allah
is the wisdom of the Shaykh, master of resignation—
free from the hallucination of personal volition,
liberated from aversion and attraction.
Impeccable in his lifelong race into Divine Love,
the Shaykh remains dauntless
in the face of the coercive forces of the world,
the barbarism of men, circus of the blind.
Humanitarian and humorous, this true human being
resembles the ancient patriarchs and matriarchs,
the shining circle of spiritual leaders
who give clear direction to humanity.
Revolutionary of Paradise Regained,
force of reconciliation, force of cohesion,
the Shaykh is the high point of ethical
and educational ideals in the long history of humanity.
Through daring action and audacious expression,
the mystic guide serves as the midwife
for human community,
global citizenship, and universal concord.

O resplendent soul who dares to plunge
into the radiant ocean of Love,
the Shaykh alone holds in his instructive hands

the key that opens the crystalline body.
This spiritual body, this corporeal light,
resembling the blue robes of the Virgin,
subsists with subtle beauty in its point of origin,
lost in timeless exultation.
Implore the master to tune the resonance of this body,
which is the vibration of your subtle being,
with his altruistic care and his courteous touch,
free from arbitrariness and hypocrisy.
Implore the master, majestic on his golden throne of Love,
to open the channels of this imperishable body,
spacious as an interior cathedral.
Cross this bridge to limitless awareness,
the full moon that neither wanes nor waxes.
Live within time while linked to timeless Reality,
free from death, the illusory executioner.

The way of the adepts of Love is not to conceptualize
but to conceive through spiritual pregnancy,
giving birth to children of Divine Love, like baby Jesus.
This is the ancient maternal mystery
that unfolds in the residence of the Beloved.
Here is the cubical chamber, black and radiant,
the nucleus of the heart, without veil or lock.
Whispering intimately the poetry of union,
the Shaykh imparts the implicit story of the universe.
With profound impact, the guide reveals
the immense net of connections known by sages,
the sheer completeness that cannot fit into words,
the eternal procession of worlds and souls from the One.
With the pen of universal intellect,
the Shaykh writes boldly on clear paper,
blank pages of the exalted heart, totally attracted to Allah.
This constitutes the interior scripture
that nothing can corrupt or stain,
confirming every historical revelation.
This is the passport to every spiritual plane

and to the homeland of Essence,
signed by our holy pastor and patron,
the universal light of religion, Nureddin Jerrahi,
who waits at the end of our immense journey.

O lover pregnant with hymns and psalms,
praise the Shaykh, the center of gravity of Love,
the uninterrupted panorama of Love,
the shining New Jerusalem of Love.
Envision within your being
the secret aspect of the Shaykh,
his essential nature, the Burning Bush,
the mysterious root by which death is overcome,
providing entry to the unitive state
of utter simplification.
Now face the Shaykh, the emissary of Truth,
who is completely free from prejudice
and liberated from the chaos of fragmentation,
who uproots sorrow from the human person
and deifies the life of humanity,
who transmits the universal patrimony—
salvation for humankind as one mystical body.

O advanced lover, stunned by Love,
whirling in the open space of the universal heart,
surrender yourself to the Shaykh,
the human channel of Divine Energy.
Be infused! Be divinized!
Steep your existence in the Shaykh.
Consume yourself in the conflagration of his love
and experience spiritual rebirth—
free from the magic cloister, the magic lantern,
the private world projected by the shadow self.
For fully liberated lovers of Love,
one unique Light resplends in myriad forms.
Before eternity, Divine Reality opened
the vast dimensions of the human heart

as an eternal image of eternal Love.
The absence of any frontier within Reality,
the freedom from subjectivity and objectivity—
this is the deification of humanity.

Inebriated, the Shaykh proclaims,
"Here and now, within inexpressible Glory,
there is neither I nor Thou.
There is no inertia, no fallacy,
no limiting individuality.
O exalted companions of Love,
mature mystics of universal Islam,
let us announce the future splendor now
in this clear space, without shadow or footstep.
I am both lover and Beloved,
both the thirst and the joy of its quenching.
I am all creeds, all sects.
My hymn to Allah proceeds only from Allah.
There is nothing but omnipotent radiance.
Essence of my being! All of my all!
I am realized humanity."

O honored guest of the universal guide,
astonished by his cosmic abode, his Ark of Noah,
your heart is a golden setting for the diamond,
receiving from the Supreme Jeweler
sweet hammer blows of *Allah Allah Allah*.
Embrace the enigma of the Shaykh,
his multiple facets, his subtle interior lights.
Become still and silent in meditation
beneath the fragrant bower of the Shaykh,
englobed by the Shaykh.
Praise him with great expansion of heart
and learn to chant as he can, with fearlessness,
la ilaha illallah—no object, no subject, only Allah.
The blood sings like quicksilver in his veins,
la ilaha illallah—there is nothing but Allah.

Experience the efficacy of his well-aimed gaze,
the sacred arrow, *la ilaha illallah.*
His mystic wedding ring sparkles in the sunlight of Love.
His royal cloak glows, embroidered with golden ecstasy.

O lover of this beautiful conqueror of hearts,
contemplate the Shaykh ceaselessly.
Intertwine your being with this incomparable being
who has enraptured the universe,
fulfilling its ancient yearning.
Come closer and closer to the Shaykh,
supremely trustworthy and worthy of esteem.
Enter his refined heart,
forged by Love from Love.
To taste his selfless welcome
is the foundation and verification of Love.
Only purified perception can interpret
the mysterious hieroglyph of the Shaykh.
Know his ardor, his impassioned touch,
and then, merged in the Shaykh,
endure with the entire force of your being
the millennia of longing of the whole creation.
With this universal longing, call out to Allah.
To recover the sense of the newness of life,
become a fish in the ocean of the Shaykh,
a star in the morning sky of the Shaykh.
Become utopia on the planet of the Shaykh,
his community where love alone reigns.
The constellations like drunken swarms of bees
fly in consternation to his sweet lips.

This beloved one of Allah—consoler of the people,
book for the illiterate, miracle for the multitude—
by teaching the mystery of femininity
softens hearts that have grown hard.
His dervishes, advanced in the way of Love,
envision all the noble envoys of Allah

within the spiritual form of the Shaykh—
within his poise, fairness, and faithfulness,
his constant absorption in purest ecstasy.

The Shaykh is the spiral stairway of light,
the progressive clarification of daily life
with its deceits, obstacles, and fictions.
The consciousness of the Shaykh,
remaining in tranquil neutrality,
is the theater where the universe unfolds.
Selected by Allah from before time and eternity
to clarify worlds, to diagnose hearts,
the Shaykh is a refuge
for the helpless, deprived, and outcast,
raising his voice in eloquent protest
against the oppressive regime of slavery to self.
His very soul is the absolution of error and negation,
a brilliant beam of Divine Providence
helping minds to overcome contradictions,
evasions, abstruse discussions,
helping hearts to reencounter primordial Unity.

The sphere of the Shaykh is without equal,
his fruitful language untranslatable
into the foreign words of the profane world.
The spiritual refinement of the Shaykh,
his fineness and sweetness,
and the beauty of his movements
demonstrate the courtesy that flows from Divine Love.
The Shaykh is salt of human existence,
stimulus for spiritual inquiry,
promise of mystic betrothal,
play between lover and beloved.
Delighting in the exuberance of Divine Love,
he is puppet of Love, jester of Love, heretic of Love.
Adopting new vestments at each instant,
he is slave of Love, sultan of Love, prophet of Love.

Most deserving among qualified persons,
firmly seated on the cathedra of human dignity,
the Shaykh is the statesman of the One Divinity.
He receives, with great circumspection,
the sincere esteem of souls and archangels.
His appearance is noble and immaculate,
stature impressive, carriage erect, eyes tender,
manners gentle, mind liberal, heart robust.
His kindness toward others is inexhaustible.
Living in neither Orient nor Occident,
abiding neither in this world nor in any other,
beyond both Paradise and hell,
the Shaykh is black space, full of constellations,
sowing the human species with stars of dervishes.

Without walls, without stopping places,
the vision and mission of the Shaykh are amazing.
Miraculous is his mosque of total coherence,
its calligraphy the channels of the subtle body.
His scimitar is the affirmation of Unity,
its jeweled scabbard Sufi poetry.
His game is chess, its checkmate mystic union.
The wings of his soul are the light of dawn.
The fragrance of his body is myrrh.
The climate of his being is subtle circumspection.
Muhammad, may peace be upon him,
is his inspiration and aspiration.
His winged word to humanity—
Rejoice! Rejoice!—
is purest gold without alloy,
the distilled wisdom of Love,
the copper of the mundane body
transmuted by the alchemy of Love.
His virtue is honesty, his faith the capacity to love,
his prayer carpet the entire planet.
His liturgy is the mystic circle,
his litany *Allah Allah Allah.*

His war to the finish is to annihilate the tyrant,
the limited self, most dangerous among dictators.
His orbit is metaphysics, the disappearance of distance,
his geometry the conch shell and the lotus.
His mantle is Prophecy, his feast day the End of Time.
His vestments are wool, his tears joy.
His wet nurse is the sacred scripture of the earth.
His consort, always at his side, alert like a lioness,
is the veracity that never ceases to inquire.
His family is all conscious beings,
his home the full spectrum of Divine Light,
his lineage the lovers of Truth, submerged in Love,
his friends the turtledove who murmurs in the morning
and the camel who crosses the desert of longing.

His only sustenance is silence.
His eyes are the eternal and the temporal,
his meditation seat the divan of white silk
in the alcove of the Beloved.
His song: "I long to remain forever near you,
for your nearness alone is sufficient, O Beloved.
As the Angel of Death reaps,
the soul prances freely on the infinite green.
I am no devotee, no ascetic, no theologian.
I am the One. I am the One."

Gazing at this bright meteor of Love,
this priceless jewel of Allah Most Transparent,
the soul of the lover, with a shivering thrill,
suddenly perceives the significance of the Shaykh.
Sinking in the ocean of Love, the dervish exclaims,
"O Allah, Your Grand Shaykh,
this explorer of all levels and dimensions,
is no sage with white hair and beard
but an abandoned child of Love,
wandering the tangled streets of Love,
lost in the Divine Labyrinth of longing and ecstasy.

He is king in appearance,
mystic orphan in reality."

Who can grasp the meaning of this vast allegory,
this secret text of the Shaykh
written in mysterious calligraphy?
Poets and thinkers remain speechless before him.
Sufi sages are almost mute,
presenting germs of Truth, mere sketches of the guide.
Even Divine Love is in search of this elusive child,
wrapped only in Light, eyes closed to all that is imperfect.
The smiling Shaykh sees himself
as a humble and unknown stepping-stone for the Beloved,
as corollary of the Beloved,
always running the risk of disappearing into the Beloved.
Leaving his heart forever open to the Beloved,
he is always dealing with the Beloved,
stirring up universal love for the Beloved,
increasing this love, linking his being with this love,
unveiling and heightening this love,
adventuring in this love, breathing this love,
without perceiving any boundary
between human love and Divine Love.

O indelible memory,
shining countenance and heart of my Shaykh.
O radiant verse of the Glorious Quran:
Wherever you turn, there is the Face of Allah.
O meeting with the Most Near.
O bath in the hot springs of Love.
O foundation of all blessing, known by his fruits.
O mysterious concealment.
O occasion for hearing and listening.
O spiritual formation of the soul.
O model for the sacred way.
O wick and oil of the precious human body.
O penetrating awareness and skillful heart.

O one who truly knows his own nature.
O one who has no likeness anywhere.
O stone basin of initiation in the sacred pine grove.
O spiral ascent through seven mystic caves.
O renunciation of limited self and limited world.
O language of Paradise.
O powerful guide with profound resources.
O general among the forces of Truth.
O well-being of humanity.
O immense good will.
O continuous prayer, source of remission.
O loyalty to the Beloved, vessel of intercession.
O cradle of faith and mother of believers.
O secret biography of Khidr, the Prophet Elijah.
O living word of the sole Creator.
O microcosm, cause, act.
O constant remembrance, science of nearness.
O cessation of the pronouns *I* and *you*.
O consecrated joy.
O snow-capped mountain.
O nest of the soul.

O delight of the three Wise Kings.
O precious pearl, green pasture, never-fading rose.
O mysterious structuring of the universe.
O clarification of the universe.
O palpitation of the one heart.
O covenant between Divinity and humanity.
O bread of life.
O genuine kernel.
O baker, potter, carpenter.
O rejoicing of the dervishes.
O happiness and sustenance of the universe.
O keel of humanity.
O lighthouse of the last port.
O magnifying glass of wisdom.
O leader and judge.

O wise countenance and firm heart.
O concord among conscious beings.
O friend of wolf and jaguar.
O shining body of Divine Energy.
O four directions of the sacred earth.
O warp and woof of existence.
O true human being.
O Shaykh.

NEW LIGHT ON SUFI SCIENCE Composed in Spanish through the method of divinely guided association, this ecstatic outpouring can barely be called an essay in the usual literary sense. Like a giant baklava, or many-layered Turkish sweet, it is delicious to taste from any direction. There is no beginning, middle, or end. Its richness precludes reading the essay in one sitting, as in the case of the preceding poem. This composition was originally intended for intellectuals of the modern and postmodern age, as a call to ignite intellect with the flame of ecstatic insight, or gnosis. There is scant reference here to the concrete details of historical Islam. This essay presents instead the pure mathematics of Sufi science, a term that denotes the organic, God-oriented structure of pure spiritual principle that constitutes the foundation of every historically revealed tradition and, in fact, every civilization. However, this *new light* shines only through the author's initiatory connection with the scientific genius of his founding Pir, Nureddin Jerrahi, whose name means Great Surgeon, Light of Universal Religion.

14
New Light on Sufi Science
Gnostic Unveiling and Awakening

> One of My servants whom I have filled with My Divine Mercy and to whom I have taught the Science that emanates from Myself...
>
> —*Holy Quran 18.65*

> Know with the Science of Certainty that tomorrow, before the Divine Gates, the seventy-two sects will be only one.
>
> —*Fariduddin Attar*

The Sufi is hidden, even from his or her own eyes—an enigmatic figure for whom Divine Mystery, the Unity that embraces everything, is always sufficient, whether in the going and coming of daily life or in the cycle of emanation and return of the entire universe. *The Only God is always sufficient to His servant.* This traditional phrase of Islam is constantly on the lips and blossoms from the very existence of the person who belongs solely to Divine Sufficiency, who remains in the life-giving atmosphere of Sufism, the warming sunlight of Sufism. Divine Sufficiency is Sufi science.

In the global civilization of the present, as in ancient epochs, Sufism is a secret, unifying, and renewing current that flows through the ocean of human consciousness, subtly touching all religions, cultures, and souls. The delicious science of Sufism, this knowledge saturated with love and wisdom, is not essentially a historical phenomenon, even though its temporal roots penetrate into every stratum of the history of prophecy. Sufism in its most internal sense is the transcendent expression of the Logos. It is the transtemporal descent of Universal Intellect—the First Light, which illumines the Original Source. Logos is simply the overflowing illumination of this sole Source, called the One God by prophetic tradition. This Light from Light constitutes the essence of human consciousness and all consciousness in the universe. Logos, always descending and facilitating the ascension of incarnated souls, is called in diverse contexts Holy Spirit, Muhammad of Light, and Interior Master. It bears all the powerful Divine Names revealed to humanity through its ancient and wise traditions. Logos, or Divine Word, is the power of manifestation belonging to the primordial Essence, which subsists beyond names, beyond languages, beyond codes, beyond structures. Sufi science is the conscious return of the rays of Logos to their shining Source.

This essay is not an attempt to describe Sufism but simply to be Sufism. According to the permutation of letters that is natural to the structure of the Arabic language and that is practiced consciously by the mystics of Judaism in Cabalistic contemplation, *sufi* means *fusion.* To know Sufism is not to sketch certain ideas but to incarnate Sufism—the experimental science of fusing subject and object, fusing human interiority and divine interiority. This process, which usually manifests within the alchemical furnace called a Sufi Order, transcends the limits of language. The distance evaporates between the object and the consciousness that contemplates the object. Only the Living One subsists and remains—the Unique, the Intimate.

This fusion of human and Divine is not an unstable mixture of distinct energies, not strange hybridization but harmonious union. The instrument of this unification, Sufi science, is a spiritual cosmopolitanism that embraces, within the Living One, the rich experience of all epochs of our long history and our future without limits, making it all totally present. To Sufism belongs the role of transmuting, harmonizing, unifying, and fusing.

The cardinal secret of Sufism can be expressed in this hermetic formula: *physiology is theology.* The precious human body demands a spiritual interpretation, an esoteric hermeneutics. The cosmos experiences accelerated evolution when human life is transmuted by the mystic love that is true wisdom, when human existence orients and submits consciously to Supreme Unity as its own inner core. The esoteric meaning of the sacred text of our subtle body is the base of the underlying conviction of Sufi science—perfect correspondence between earth and Paradise, between sensation and supra-sensation. Free from sectarian dogmatism, Sufism is not an abstract conceptuality but an intimate knowing, the fruit of direct experience, acquired through various modes of spiritual deepening and gnostic unveiling. This Sufi way of knowing ends in illuminative and transformative union, final goal of the arcane path of purification, the ancient road of Love and Wisdom.

Our human experience, elevated to its highest level during the thousand and one luminous nights of mystical practice, culminates when the creature becomes transfigured by the unique Light of Essence. The fundamental principle of Sufi science is the aptitude of the human spirit to participate totally in Divine Essence. Logically and theologically impossible, this wedding between finite and infinite allows the aspirant to abandon the limited self completely in order to enter essential Reality. This can occur only through the great ecstasy that lifts one beyond the conventional invocation of some divinity who remains separate or distant. Such ecstasy of insight liberates us from the ambiguous or conflictive situations that we illusorily identify as our own humanity. Such ecstasy unveils Essence.

The entering into essentiality called Sufism is not a solo flight into the heavens. It is a beautiful, affectionate, kind manner of behaving with others in this world, in this moment—without depreciating or underestimating any living being, without encountering any wall of separation. Sufism is the aspiration to be essential, to possess entirety of character, inner coherence capable of casting revelatory light upon everything that exists, gradually unveiling the true nature of everything. This is how Sufi science reveals the unique Essence of the innumerable spheres of meaning, which are successive initiations displayed by the One. This transcendent science does not negate natural phenomena but simply knows that meanings are rooted not outside but always inside the One. The plenitude of Being contemplated by this science excludes all lack of meaning, all sense of profanation or impurity, all failure of communication among the infinite dimensions of Being. Sufism refuses nothing, with the exception of the empty pretension of being something or someone independent or distinct from the Living One.

O lover who listens to primordial silence! O swan who parts the smooth waters of Essence! What happiness flows from the soul who becomes consciously capable of a love totally unified with the One God, fully absorbed in the Most Near, Who is Love! Manifesting beyond the duality of humankind and Deity, beyond surface teachings and exercises of religion, this state of total absorption actually constitutes Sufism.

Divine Sufficiency is Sufi science. Along the mystical path toward realizing this all-embracing, all-conscious Sufficiency, it is necessary to pass through an initiatory process, submitting to various spiritual tests, which arrive spontaneously and unexpectedly. The experienced direction of a masterful guide, pivot of a mystical fraternity, is usually necessary. One must sincerely adopt certain modes of behaving and being, the gestures of a contemplative Order that spins around the center of its saints and sages, the spiritual sultans of its initiatory lineage. To submit to the direction and the correction of a guide and to complete a period of novitiate in an Order that has survived the vicissitudes of history, is not mere adherence to an ideological group. It implies ingression and integration into a God-filled body, which does not simply think about God but lives God.

Strengthened by the subtle interior influence of his or her

community, the one who has come into contact with the Order now decides, out of new personal maturity, to follow the mystic path. Novice is transformed into adept during the ceremony of enthronement, the initiation into Sufi science. This spiritual journey is not the fruit of isolated individual efforts but of mutual spiritual affinities. The flow of Essence, called Divine Grace, propels this movement of the soul, which culminates in illumination—conscious union with God, the intimate merging with the Intimate, the inward with the Inward. The disciple of Truth, filled with enthusiasm, realizes and becomes the Divine Word. This realization constitutes the force of Sufism. The authentic practitioner of this science of Divine Sufficiency is lifted above the doctors of religious law who often persecute Sufis, from whatever tradition, judging them to be heretics.

The inert world of norms and crystallized habits exists as a tense equilibrium between the superficiality of superstitious or rational religion, on the one hand, and the conventional pessimism of a basically atheistic society, on the other. This archetypal situation, partly humorous and partly tragic, does not affect Sufism, which is lived in interiority, far beyond external appearances. Sufism is the call to penetrate through appearances. It would be useless to look for Sufism within the sociocultural framework of any religious tradition.

Sufism is the invisible college of spirits liberated and illuminated in God. It is a secret comradeship on earth, as well as in Paradise and in the Garden of Essence, an intimate friendship among advanced practitioners and sublime masters. This parliament of gnostic saints and sages definitely exists, quite apart from what is commonly thought or imagined. It is the obverse of conventional human community. It is the meeting and union of temporality with eternity. This apostolic college on earth stimulates transformations within the religious consciousness of the epoch, transformations that have repercussions as well in the circles of political and social power. However strange it may seem to the modern mind, all peoples on the planet, without being completely conscious of it, maintain mysterious contact with the masters of interior perfection and with their advanced disciples. This mystical body of Sufism is the universal intercessor.

One could describe Sufi science as the faithful interpretation of all books of revelation from the labyrinthine history of humanity. But formulas always falsify. Sufism is not the interpretation but the origin, the illumined love and ecstatic wisdom that overflow the limits of all categories of thought, including the notion of history. This essential science can communicate without the mediation of images, scriptures, doctrines, analogies. Forms of classification tend to separate, and this science is unifying. To comprehend the Sufi message—to decipher this code, to handle this master key—demands great subtlety, a marriage of exultation and sobriety. Sufism is the fundamental union, the common nature, the consubstantiality between Divine and human. In Sufi science, Reality alone acts, penetrating veils and rendering them transparent. Our own most intimate gaze is the gaze of God.

We assimilate Sufi science according to the plane upon which we focus our attention, our insight, our longing. This process of assimilation with its charismatic signs of change, which are preludes to a more radical renewal of our being, this maturing through a spiritual adolescence, transforms daily experience into timeless knowledge, filling every moment with the savor of eternity. Tension disappears between anxiety to attain a world beyond this world and acceptance of our condition on the earth, for this present condition is now revealed as an expression of the One, as transparency to the One. We abide as essentiality. All ambiguity, in the widest sense of the term, now disappears. Our false image of this world evaporates. It is no longer perceived as a landscape of hallucination—a purgatory of separation from God, of continual failure in our intention to unify consciousness. Daily life now becomes the ceremonial dance of Sufism, celebrating the reconciliation of heaven and earth. Everything reveals its essential face. We experience the manifestation of a new light, a new mode of existence.

The expulsion from Paradise is not merely a symbol for our human condition. This original fall does not exist in the past. It is the persistent clamor of our limited self, which negates the dignity and ignores the necessities of others, which rejects the Paradise tenderness that is based in our union with Divine Love. Similarly, the mystical return to Eden is also not just symbolic but actual, immediate, experiential. No other experience possesses the same intensity. The ordinary world reveals itself, suddenly, as something never encountered before. Without superficial emotion, we hear and see in an entirely different modality. Everything becomes lightweight and turns transparent to the One. This is the magnificent sovereignty of the soul's consciousness.

In the Garden of Essence, which embraces both Paradise and the earthly plane, one breathes an oxygen rich and dense. The world of noise and turbulence dissipates like clouds. This is the mysterious rediscovery of our spiritual body. In an instant is revealed the Divine Sufficiency, which we have gradually forgotten. This is the supreme spiritual healing, the complete resurrection. We are amazed and bewildered before the unspeakable. We see the original face of our eternal soul, which is a ray from the One. This is not some form of pantheism, because no separate material medium exists through which the Divine Energy is diffused. Nor do there exist in the Only Reality any of the supposed divine strategies conceived by men. This Paradise of here and now, discovered anew thanks to Sufi science, is a pure astonishment, a total and almost insupportable vision, a fervor without object or subject. It is the complete opening of the heart.

Paradise experience, this epiphany of the infinite through the finite, is the key to Sufi science. With great initiatory intensity, our basic identity comes into question. This unexpected reunion of the soul with its own essentiality can stimulate the entire spectrum of sensation from existential dread to transcendental radiance. But cowardice is consumed in the flames of ecstasy. Those who have been destined, who have been illumined, who have been anointed as humble servants of humanity are transformed into instruments of God, now completely cleared of the ephemeral self. Only in this very elevated spiritual state—pure thanksgiving and pure adoration—can the disciples of Divine Love wear the cloak of wool, the traditional sign within Sufism, the holy lineage of Muhammad, that signals an attitude of true detachment, a perfect abandonment of the limited self and its limited world. This is the renunciation of the self in God, the losing of the world in God—the blessed annihilation that blossoms as resurrection.

With the supernatural power of its wisdom, Sufi science penetrates to the interior of sheer existence and discovers itself face to face with the unique Essence. In this manner, Sufism attains release from the labyrinth of mirrors, the way of living that is always unsatisfied, always separated from Reality. The life of self-concern eventually leads to moral solipsism, to the negation of humanity. By means of transfiguring ecstasy and spontaneous altruism, this science of selfless consciousness teaches the spiritual way of interdependence—interpenetration among all souls, among all conscious creatures, among all cultures, among all the living spheres of being. This rigorous science does not simply reveal the dynamic coexistence of everything but also unveils the absolute identity of everything, affirming without hesitation the most radical principle of unity.

The universal reconciliation this science effects, which is a mystical fusion beyond perception and beyond intellect, does not imply any reduction of the sacred singularity of each being. It involves a radical transposition but not an extinction of our own being, our genuine personhood here and now. Sufi science imparts a subtle but decisive turn to our entire existence. Human beings continue as human beings, exercising their precious free will, the trait that is most characteristic of our humanity, that is the inalienable right of our being.

With its purificatory energy, this spiritual maturation called Sufism truly restores the dignity of humanity. The social world, including all the fanaticism and intolerance that manifest in every historical period, finds itself englobed by the secret force of Sufism, which is the animating substance of civilization. In this search for the Holy Grail among the Sufis, no insuperable obstacle exists. Thanks to the illumination provided by Sufi science, we become conscious of never existing except for and through the unique Reality. The spiritual combat against Satan, the Perverter, is recognized as the combat against our own fantasies, which are very dangerous but which lack any substantial reality.

One must constantly begin anew the common task of Sufism. We can never shrug our shoulders fatalistically. The way of valor called Sufi science is not constituted by effusions of emotion, by vertiginous successions of dreams and visions, or by the exhibition of miracles. The spiritual path is upright and direct, remaining open simply for the next step, simply for each breath, simply for the unbreakable commitment to serve humanity, simply for the patience and courtesy of Sufi behavior. The cycle of human effort and Divine Grace is never closed. In each place, in every moment, we have to begin anew the purification of the self and the humble service of all creation. This constantly beginning anew, through the continuous impulse of love for God and love for every creature in God, constitutes the steps of the sacred dance of Sufism, accompanied by the rhythmic beat of the universal heart. In this dance without description, the Divine Will reveals a cascade of surprises. This inexhaustible Divine Manifestation becomes our path to conscious harmony with Essence.

Without an immediate acceptance of Divine Will, our perception and comprehension of life remains partial, loaded down with passionate impulses and instincts. Without conscious immersion in God, the universe appears as multiple series of fleeting forms like fireworks, either meaningless or interpreted by egocentric expectation and speculation. One must hear and see instead with senses given over entirely to Essence. Then the Divine Creation displays all its richness and harmony—as a dynamic unity in which contraries are embraced, as a perpetually unique flow of transparent creativity, and as a mystical flight of the One to the One through translucent veils of multiplicity.

Gradually, without our reasoning about it, we encounter the unknown, the ineffable, expressing itself directly through what is concrete, what is palpable. This advent of direct perception and comprehension of Essence is the charisma of Sufi science. What is generic or abstract disappears into pure Presence. It is no longer necessary to distance oneself from the world, to retire into a hermitage. The engaged Sufi lives always contemplating not the reflected universe but the essential plane of Divine Love. Every detail of existence manifests through and as Love.

Speaking the utterly unspeakable, Sufi science is neither a description nor a history but a song or an absorbing vision without fixed images. It is not an intellectual theory but an enchanted world where only the power of the Word prevails, only the virtuosity of the breath. In this central science, which is the point of intersection for all worlds, frontiers open among distinct disciplines of knowledge. Sufi science is the organic whole, composed by all forms of knowledge, all currents of experience—a place of germination where the embryo of the new being incubates. But this science is no subterranean cave. In the wide open space of Sufi science, a transparent and resonant atmosphere always prevails. We feel equidistant from all extremes. Without making any effort to provoke it, we give free rein to the natural tendency of contemplation, latent in each person. We wander and play in Divine Beauty, where Sufism blossoms, leaving thought and feeling free like a tide of praise that ebbs and flows.

Through the intellect illuminated by Divine Light, Sufi science transforms and elevates the ordinary disciplines of religion. Touched by the ecstatic flame of Sufism, forms of religious obligation lose their anemic, habitual meaning, melt, and flow, animated by a new life, purified of dross by a new light. This holy and ecstatic religious freedom, suffering no lack of inner discipline, is an unequivocal sign of the spiritual vanguard called

Sufism. Beneath all authentic religious forms, Sufism is hidden. Whenever a great Sufi manifests, he or she is always ahead of the times. The perennial religion is then born anew in a jubilation purified from guilt and anguish, purified from automatism, purified from autocracy. The periodic surging forth of Sufi masters from every religious tradition punctuates the spiritual development of humanity like the drumbeat of an ecstatic heart.

Sufi science expresses in a rigorous manner, with totally awakened consciousness, this most ancient affirmation: *Reality is One*. The mystical interpreters of all sacred scriptures, those whose gaze penetrates deeper than the visible world, always discover this supreme treasure of Unity, this Divine Peace in which all enmity and rivalry between different points of view ceases. To abide consciously as the One does not mean to take refuge in some imaginary childhood paradise or prenatal paradise, nor in the Paradise beyond this world, as sublime as it may be. To be the One means to stand here and now in the omnipresence of Essence—without any sense of exile or dualistic perception.

This awareness of subsisting as the heart of the One, the Reality that is hidden only on account of its excessive transparency, is not a static attainment but a continuous beginning. Even though everything is experienced as one and the same, the functional character of the world does not disappear but manifests openly as Divine Will, as the emanation of the One, always giving birthless birth, always expressing its Beauty, Mercy, and Power.

This science of the One is not transmittable through formulas of faith, reasonings of philosophy, or techniques of meditation. Sufism is the ancient path never forgotten by true adepts and their disciples in each generation, who attain a mysterious fine-tuning between body and spirit. Following the counsels of ancient sages through continuous Oral Tradition, they rediscover and verify the mystic correspondence and final equivalence between the human and the Divine. Sufism celebrates an alchemical wedding that is incommunicable. It is an esoteric affair of points of energy and confluences of spiritual currents within the subtle body of humanity, which is the perfect microcosm.

Wise and tender Sufis teach without words this wordless doctrine of the One, using as instruments of transmission the luminous dreams of the aspirants, interior revelations, which contain an infinite possibility of interpretation at various levels. This rich plurality of meanings never disappears. The unifying operation of wisdom always remains multidimensional. The Living One is not comprised by any linear structure but by organic orders of parallel realities in which there exist, blossoming from the sole Consciousness, numberless unexpected versions of the Real. This limitless profusion of meanings excludes only nonsense, or absence of meaning. The amazing life of the One has no relation to any mathematical singularity. The One is the teaching of Sufism.

This deep spontaneous teaching—this creative participation in Divine Wisdom, this silent communion with Divine Unity—does not represent but directly presents the One. It is an immediate perception of God, Who is immanent in the depths of the human being. It is the recovery of our original richness, our primary nature of timeless and limitless awareness. The very same consciousness belongs to conscious creatures in countless universes, who all live by centering themselves in God, the Creator and the One. The complete experience of this consciousness is what Sufism means by the culmination of revelation, the end of the cycle of prophecy, the goal of humanity's journey from opacity to transparency, from phantasmagoric dream to panoramic clarity.

The elected ones in this noble chain of initiatory succession called Sufism themselves constitute a mystical exegesis of the universe, embracing all living beings as most intimate neighbors, as rays from the One. Sufism is not an elitist doctrine. The same Divine Light acquires a distinctly nuanced quality through each individual human consciousness in diverse cultural environments throughout the course of millennia. Through this mysterious process of Divine Manifestation, Oneness sketches the contour of

Its infinite possibilities. Each facet of this sole Consciousness reflects all others. This resplendent web of reflections is not a passive image in an inert mirror but a mutual current that crosses over every subjective and objective frontier. The whole life of creation is the pilgrimage of the One to the One. All desire is the spiritual longing of the One.

The mode of behavior observed by the adepts of Sufism is never capricious or arbitrary. Manifesting with surprising differences and similarities produced by vast cultural interchange, Sufi discipline maintains always precisely the same compassion and courtesy, thanks to the impulse generated by the same ardent love for God. The reciprocal sensibility, the continuous mutual affection, the unique attraction of love, the mystery of union between lover and Beloved—this is what really counts in Sufism. Universal friendship is the common note that unifies Sufis, who become the invisible lineage that links and bonds humanity in Divine Friendship with all Prophets and their holy companions throughout time and eternity.

The figures in this glorious prophetic lineage who stand out with greatest clarity for Sufism are Jesus and Muhammad, may Divine Peace always embrace them. The beloved Jesus and the beloved Muhammad are manifestations of Divine Awareness. They are beings avid only for God. They incarnate Divine Love and Divine Wisdom. They demonstrate and continue to transmit disinterested love for Truth. Maintaining great reverence for their noble prophetic precursors in all the epochs of human history, Sufi science, avoiding dogmatic definitions, concentrates in the blessed contemplation of these two sublime beings of love and wisdom. Each with his own spiritual mood and tone, separated in mundane time by a brief period of five centuries, Jesus and Muhammad mark a new step in human history—the birth of our spiritual era. They miraculously inaugurate the final days of completion, which will have an indefinite duration—a future of a split second or millions of years, comparable to the geological history of the earth. Only God knows.

These final days of rapture and plenitude that now surround us are an immense rhapsody, a passionate lyrical sigh, a profusion of scintillating revelations. This is a perception of theophanic transparency much more profound than the viewpoint of any empirical historian. From its first witnesses in prehistory to its realization in the secret profundities of the vast planetary communities of Jesus and Muhammad, this maturation of the human spirit has been channeled in every cultural ambience through Sufis, who have always stimulated rich reciprocal influence among historical religions.

To return up the current of Being, singing lyrical songs of love, has been Sufism, is Sufism, and will be Sufism—the spontaneous movement that existed in the great classical civilizations of the globe and continues to exist in the modern world, where a new generation of Sufis is emerging, singing their hymns of infinite joy and audacious aspiration for all humanity. Sufism is the serious invitation from Truth for us to take responsibility, with entire good faith, for working out the well-being of humankind and the planet. The life of the Sufi, in whatever context, is validated by the experience of total affection for humanity, total consecration to all creation, and conscious unity with the Living One.

Humanity left to itself, its life drained of spiritual significance, follows the false doctrine of the survival of the fittest, represses its painful sense of exile, and enters gradually and unconsciously into darkness. Lacking a true center of gravity, human beings then experience upon this earth the eruption of the Inferno, with consequent moral devastation. Sufism is the alternative current in history, even the secret essence of history. Sufi science, which is disconcerting to those who are attached to rational dialectic, is an unveiling of Divine Grace, which exists everywhere, always, in all hearts. This subtle science, which refuses to incarcerate the Holy Spirit in formulas, calls humanity to enter consciously the field of universal brotherhood and sisterhood and to look upon physical nature as an open book of spiritual symbols. Sufism plants its inspiring message in hearts full of nostalgia for revelation—for a true paradigm, for a spiritual language that could replace the analytic, technical, didactic, and manipulative language of the world. The living words of a Sufi master, who is devoted passionately to the service of Divine Beauty, must be of an order distinct from all theology and philosophy, must be the opposite pole of all ideology.

Sufism generates not ideas but luminous waves in the ocean of human consciousness, waves that are regulated by a secret rhythm. Sufism is the interior regenerative force that succeeds in elevating the entire human enterprise, inspiring authentic efforts at purification and persevering meditation. Sufi science is not a magical activity. Magic is always the quest for power, the cult of power, the drug of power—whether expressed through ancient atavistic rites or through modern technology. The stealing of fire, the Promethean rebellion, is equally the drama of contemporary society. As illuminating as the sacred history of our lost Eden, this theft of raw power by the ego represents any action that tends to destroy our direct sense of the unity of consciousness. It is not simply a childish myth. It is the one sacrilege possible: to negate unity. Sufism is extremely sensitive to the danger of such negation, which may come from any direction.

Sufi science is simply the affirmation of Divine Sufficiency, the indivisible plenitude of the One. This science, free from anxious grasping for personal or collective power, does not operate like a therapeutic doctrine or a religious strategy for redemption, but effects a fundamental fusion with Divine Life. Sufism does not interpose intermediaries. It is direct vision. In its intuitive totality, nothing is suppressed and nothing is abstracted. It is not an ideal outside ourselves, nor even our organ of perception. Sufism is simply to experience the perfect forgetfulness of ourselves. It is a visionary activity that spontaneously intuits the secret relations among numbers, letters, dreams, stars, and daily events. It provides telepathic communication with living beings and those who are called dead. It offers the gift of premonition and the gift of deciphering the universe as a play of signs and omens. Sufi science is clear, naked vision, before which is displayed the mystical itinerary of the soul, stage by stage, each marked by precise degrees of love and wisdom.

The panoramic vision of Sufism unifies epochs and cultures into an essential expanse that embraces human life in its totality, in its transparency to the One. Our ordinary concepts of space and time must undergo radical transmutation by Sufi science. All conscious beings witness and testify to this englobing presence of Pure Life, sometimes called Divine Presence. The Beyond manifests as the Near. There is no other being than Being.

The actions and words that affirm the Living One surge naturally from the heart of Sufism like flower from stem, engendering in the practitioners a disposition of mind capable of transcending accepted and codified conventions, capable of transcending mere articles of faith, labels, and titles. Instead of being a system of intellectual coordinates, Sufi science displays its principles through dance and pantomime, through playful and profound gestures of the Sufi master and his or her community. Sufism is the organic rite of steps and breaths that mysteriously actualizes the principles of spiritual science in the human body, the rite of the mystic circle that celebrates, here and now, the new cycle of wisdom, the new light. Sufi science always teaches the accompaniment of prayer and contemplation with beautiful and subtle movements of the body. It is not a medium of knowledge but a medium of metamorphosis. Far from being rarefied and abstract, Sufism displays itself not as a rational scheme but as the concrete dance of all existence immersed in God, radiated by the One. From childhood, veneration is natural to human beings. In Sufism, this primordial urge to venerate, which is intrinsic to the soul and is perpetually being displayed in various forms, finally expresses and incarnates fully, not within a mythological paradise or inside a cloistered convent but across the face of the whole earth.

The kingdoms and governments of this world exist in a permanent crisis of spiritual meaning. The contemporary myth of modernity and its supposed sufficiency is beginning to lose vitality. Sufism is a ray of Divine Light, ungraspable in itself, that transmutes whatever it touches, submitting to examination the pretensions of every society, the presuppositions of every science, all of which are incapable of fully embracing Reality. Political ideology and religious dogmatism generate the worst confusion in this universal crisis of spiritual significance. Sufism is the password that opens the gates of authentic knowing. It is not one more system of signs or conventional beliefs that we can accept or reject. Without this living keyword of Unity, this Divine Word, civilization in its interior depths lacks true foundation. Sufism is the alliance or covenant between God and the human heart, which is always in search of sure guidance. Sufism is the promise and actualization of our interior perfection.

The community that succeeds in keeping its spiritual knowledge intact, not simply as an ancient religion or as a cultural tradition but as an indivisible totality of being and as a perpetual thirst for this totality—such a community expresses a total way of life, its collective psyche dwelling in a blessed state of receptivity to the One. The spiritual enterprise of such a community, gifted with archetypal values and speaking an idiom of initiates, penetrates the various strata of consciousness and arrives at Essence. This triumph of faithfulness to true human nature inaugurates an epoch of splendor, whether it is simply in the breast of a Sufi master, in the protected precincts of a mystical fraternity, or openly as a planetary civilization.

Sufi science effects a mediation between any particular society and the deep ethical and spiritual postulates that ground that society. Sufism heals the collective deafness to the prophetic call by generating personal vigilance, which is irreplaceable and cannot be substituted for by anything—the vigilance of each human being who desires to live in communion with the One. A process of call and response always exists between Sufism and society. There is an inseparable correlation between the mystic path and the life of the people. Sufi science is the universal tradition—nuanced by lights of Judaism, Christianity, Islam, Hinduism, Buddhism, and indigenous traditions—the universal tradition where pilgrims on their long journey encounter food and shelter, spaces to pray and meditate. It is a caravansary of tenderness and guidance.

Sufi science opens a harmonious zone in our being. This science generates a state of profound release that is not hypnosis but alert introspection—free from the resistance of what is inert in psyche and culture, free from subtle forms of narrow self-satisfaction. This science investigates the innate experience of universal harmony, perceived according to the personal sensibility and particular genius of each individual as a ray from the One. In the heart of every creature, with its unrepeatable and unique character, in the nucleus of each expression of the Living One, reigns the same *going beyond*, the same longing for mystic union or completion. Sufism presents the potential open to all human beings—whatever their temperament or turn of mind—the possibility of tasting and enjoying union and supreme identity.

Sufism is the impulse toward the essential, toward the secret point of intersection with Oneness, where the human being refuses to be a mere cipher or concept and recovers its original nature as the Living One. This is the delightful liberty of Pure Being, the rapture that accompanies the Sufi pilgrim wherever his or her route may lead. It is not by chance that Sufi masters speak of liberating the human soul. This regenerative, transmuting operation of Sufi science through nonverbal forms of communication reveals souls as amphibious organisms, free to swim or dance with their holy compatriots, at once in temporality and in eternity, basing themselves solely in God.

These teachings, these exceptional gifts that transform devotees who gather around Sufi masters, do not contain any technical or mechanical manipulations, neither of mind nor of body, nor can these teachings be reduced to any planned series of spiritual exercises. On the contrary, the authentic transmission of the living principles of this science is rooted in the evocative and lyrical capacity of the Sufi master and in the capacity of disciples to tolerate ambiguity of meaning, intuitively understanding various subtle allusions that evoke a rich plurality of hidden teachings. The disciples must be able to sense their unique affinity with the soul, mind, and body of their beloved guide and the entire spiritual community, including mystic saints of the past and the common basis of their different personalities and their different epochs. Sufism is not something exterior but is the singular gaze that is shared by master and disciple in the transparency of the One.

Sufism is a radical shock that produces the interior manifestation of the long desired and astonishing Day of Days, opening the doors of ecstatic abandon to perpetual communion with the primordial purity of the One. We are drawn forth from the sinuous corridors of the limited self, from the nightmare of a moribund society. To tear away all masks, the theatrical representations of our personal and collective history, is an operation capable of fundamentally changing the world. The plethora of deceptive forms will dissipate like clouds. The prisons of our conditioned concepts will break open like seed pods.

The white conch shell of Sufism resounds with the oceanic music of the One, where divisive time and space cease to function, ceding to Original Unity. Sufi science renders navigable the Oneness without characteristics, path, or surface. This noble and sublime science, which nourishes humanity only with the bread of Truth, is based on direct apprehension of the sacred procession of the Logos within the One. This mystic bridal procession is a luminous process disguised as a material universe, a timeless process covered over with the fairy tales of ancient mythology and modern science. It is the mystery of the Universal Mass in which the Divine Word perpetually descends.

Unity alone exists—impossible to split, without spatial or temporal measure, without the slightest fissure. The beautiful procession of worlds without number within the One never ceases to coincide with the flow of the Divine Will. This creative and continuous emanation is already established in eternity. This mystery, springing up before the illumined eyes of the heart, opens them to the sacred precinct of Essence, where all phenomena melt in the sport of the One.

It is impossible to imagine with what tender solicitude Divine Grace protects souls from being dispersed in this bewildering playfulness of Reality. This Grace alone makes possible the vertiginous transition between Essence and Divine Creation, where the same Light mysteriously acts a double role: motionlessness and movement. There is simply no analogy for such unitive play. All description of this mystery—Unity that manifests as dynamic multiplicity without ceasing to be the One—remains the stammering of poetry or the empty disquisition of metaphysics. Reality cannot be reduced to intoxicated words or sober formulas. Reality continues to be only and simply Reality. Nevertheless, the Ineffable lives fully, with efficacy and power, as each one of the conscious participants in this dream of Divine Manifestation. In Sufi science, this principle of the mystical participation of each person as the One appears in full light and demands of the Sufi a life of complete originality and attention, alert compassion and humility toward all living beings.

Sufi science, diverse as its historical roots may appear, is the consistent response to our demand for universality, a response far beyond any attempts at synthesis or syncretism. The planetary community of Sufism, invisible even to itself, accepts responsibility for the urgent and saving task to discover the unique foundation of all moral principles, all religious doctrines, all philosophical insights. It is necessary to plant this task in the consciousness of the entire earth. Sufism sees the necessity of a vast reaccommodation of peoples, cultures, religions—the necessity of reawakening ancient, sacred links with the Promised Land, not through mere ritual repetitions or external talismans but through a true combat against the divisive forces of the limited self. This spiritual warfare demands a comportment of courtesy and equanimity from the submitted subjects of the mystical King who are doing battle, using the weapon of perennial wisdom against the distortions of the age.

Sufi science is not philosophy but the passionate attempt to found upon this earth the community that religions hope to encounter on the other side of death. Sufis aspire to open the floodgates of the Divine Current in the human heart on a world scale. The life of integrating into this friendship without boundaries, into this sacred common work, demands the sensibility that spontaneously submits to the One and thereby unifies with other beings as the infinite family of Consciousness. Just the birth within humankind does not grant a person his or her true humanity, which is to see Truth face to face.

Sufism is an initiatory test, an exodus, a distancing from cultural and personal rigidity. Rites of initiation and their purifying and liberating character stand forth with clarity in the history of all ancient traditions. The apparent absence of the mystery of initiation and passage is a modern anomaly. As initiates who truly live, speak, and love, we become one with our environment, with the marvelous Reality that surrounds us and constitutes us. Our initiatory act affects the whole human collectivity. Our awakened energy does not pass unperceived in the world of the relations of all living beings.

Mystical Orders that inherit a continuous initiatory lineage, whose members walk beneath the green banner of a fully realized leader and guide, are a function within the very heart of humanity, a function that always generates new revelatory light. Thus the emergence of such Orders is spontaneous and inevitable, like the pumping action of the physical heart. Spiritual life in community teaches universal compassion, the distinctive note of a contemplative enterprise that does not live to ascend to heaven but to deepen roots in the sacred earth. Every Sufi initiate recognizes that planetary community is our central responsibility, as we renew our primordial covenant with God to return to our origin, to live within Divine Unity. The apex of civilization, which is the community of the wise lovers of Supreme Love, is always situated in an essential present, not existing merely in a remote future or as a vestige of a golden past. Sufism does not symbolize or prefigure but incarnates the Reality with which its science deals. Sufism pronounces the powerful word that puts into movement on this earth the communal task of liberation—a spiritual initiative that inevitably faces the full resistance of the established environment, the resistance of arbitrary convention and virulent ego. The living word of Sufism always illuminates this decisive moment: the conscious birth of universal humanity.

In spite of our being surrounded every day by a world of tangled and neurotic emotions, a world of mirrors and echoes, this revolutionary aspiration can reign in our spirit. The call of Sufism to clarity of action opens our breast with a gnostic incision, awakening great longing to know ourselves again as a single consciousness diversified into numerous branches. This call stirs enthusiasm to follow with certainty the holy steps of our principal inspirers—the sages, saints, and Messengers who possess and demonstrate human authenticity, who never remove themselves from the sacred communal existence of living beings. Sufism forbids only passivity, abolishing by the potent decree of aspiration all that weakens our internal spiritual drive.

Sufi science always faces the universal situation, which all human beings, all cultures, and all epochs and ages share. Our false sense of alienation from the Unique Reality, our isolated acts and concepts, must be replaced by total experience. This demands a profession of total faith—a circle without circumference, a consciousness that embraces everything. This is the embrace of Essence. This is the faith beyond schisms, professed by Sufism with rigor and without wavering. This all-embracing faith displays an emphasis that is not metaphysical but existential: the development of a truly just and free society, governed and guided by the eternal liberty of the soul. It will be a surprising society, which forgets its own self-interest to love only God. It will be a society that arises at midnight and with the dawn to pray. It will be a society that seeks out spiritual masters, a society that continues to meditate in spite of privations and tests, that attempts to penetrate the spiritual sense of the universe, that discerns with clarity the causes of the decline and disappearance of previous civilizations. This is not utopian speculation. The Sufi is a being who is extremely sensitive to problems and deceptions of the present time. The Sufi is not a person who maintains aloofness. The Sufi never renounces our origin, which is humanity, or our native land, which is this earth. Sufi science is animated by the pure, inexhaustible longing to unify the entire planet in consciousness of the One.

This ethical and gnostic science is a crucible within which religions, values, and disciplines are heated, mixed, and fused, revealing new facts. This science sharpens the instruments of perception and intellection, refining our sensibilities on all levels. In this state of refinement, the newness always characteristic of Divine Creation springs up before the eyes of the heart. In a sudden illumination, historical times melt, without vanishing, into timeless Presence. We do not abandon the ancient sages, nor the genuine spiritual guides of all noble traditions, but encounter them directly in the essentiality of the present. This science of glory and delicacy offers a global solution that defies logic, demanding a series of unforeseeable changes and disconcerting turnabouts in the interior of the human being. The characteristic nature of the conventional world and the conventional self is resistance to change. Ecstasy is the efficacious way, the spiritual medicine employed by Sufi masters of all epochs to combat this resistance, this inertia, this ineptitude that, far from giving up, grows in intensity year by year.

This warrior science of the transformation of character does not just exist at the margin of history. In the form of revolutionary waves, which are always new, it periodically inundates the arid human mind, tyrannized by its autocratic ego. The components of this tyranny in every society and in every personal career are always the same and always pose problems that are difficult to resolve. Our search for a fundamental solution will be successful if we follow the mystic path to its goal—the intimacy of a union of total love, in which God melts into the human lover. Then there exists only the Divine *I Am,* replacing the idolatry of ego.

Sufi science, never anti-traditional, does not counsel humanity to abandon the powerful ancient methods: prayer and meditation, service to the neighbor, study of scriptures and biographies of saints. What is distinct about this spiritual science, which always develops anew in every historical period, is the wideness of its field of vision and activity. Sufism is never limited to a single sensibility, to a single society, or even to a single religious tradition, but rather embraces the world as a sphere without cultural or personal frontiers. Sufism is capable of inventing new words, but in each epoch its point of departure is exactly the same: to gaze face to face upon Reality. Its loom is always the same: the Living One. Its call to humanity is always the same: Let us begin to exist truly!

This primal science, without erasing the specific characteristics of authentic traditions, avoids every form of sectarianism. It breaks the deception of any religion that has become reduced to a caricature of love, to a mere rhetorical or literary exercise, to a petrified cultural image capable of imprisoning our most intimate internal being. The primary procedure appropriate to this science is not analysis or critique. As fetuses in the maternal womb of Sufism, we submerge ourselves in the Paradise sensation of praising, breathing, being astonished. Yet this wonderful prenatal life of the disciple of Divine Love is not yet mature mystic union. We must advance with more decisiveness, with courageous resolution, ignoring the vestiges of our personal past, ignoring even our own spiritual experiences. We must learn to magnetize all our actions, both voluntary and instinctive, with the longing to come forth from the chrysalis of the limited self, the limited society, the limited family, the limited religion. We must transcend false notions of Divine Punishment, whether internalized or externalized. We must enter blessed concordance with all humanity, calming our uncomfortable reaction to the tests and purifications that present themselves as human life upon the earth. The spiritual art of living within Divine Love and Divine Wisdom is an art that we will never completely master without disappearing into the One and coming forth again as Love and Wisdom.

The germinator of our spiritual seed, the principal agent of our transformation into living testaments, is the Sufi master. He or she is the axis of our evolution, the humble link in the transmission of Divine Energy, which has flowed into the human heart since the age of Adam, first human being and first Prophet, radiant with the unequivocal light of revelation. Throughout the long quest of humanity, this veridical transmission, with all its interdependent branches, has never been interrupted, even though it remains hidden beneath various masks, various dogmatic labels. Spiritual life has been expressed across the spectrum of possibility, from armed struggle to monastic renunciation. It is a self-purifying stream.

Thanks to its abundant spiritual inheritance, humanity can live without fear in what is essential, no matter what the circumstances may be. Humanity can live in its own natural atmosphere, its essentiality, like a bird in the sky or an innocent being in a Paradise age. On many occasions of our personal existence as well as throughout social history, however, this truly human inheritance, this necessary nourishment, is entirely forgotten or willfully suppressed. Yet in the most distant corners of the world, the most unknown zones of the heart, Sufi science always rediscovers living traces of this primordial inheritance, this openness to the One. The heart is always capable of catching fire in liberating ecstasy, igniting others with the sacred fire of illumination. With an ease that has no counterpart in the world of separation, the Sufi master breaks open our personal and social prison cells so we can consciously breathe the free air of the One, the atmosphere of the Only Life, transmitting the open secret of Unity to his or her spiritual successors as a living flame.

Sufism exists outside classrooms. It is an irresistible movement of our being. It cannot be reduced to any formula. It provides direct access to the marrow of experience, which we can always use for the most concrete and immediate goal: the liberation of humanity and the organic solution to its perplexity. Sufism is not an inert form of behavior, nor a fixed image of the world, nor a particular program of reform. Putting between parentheses all presuppositions of the personality, all commonplaces of society, Sufism presents an education in Divine Subtlety. This education generates an activity of our total being, free from ordinary ambitions, not because it disdains them but because it transcends them. This ardent and hopeful education is the keystone of Sufi science, which does not submit Reality to any rational or emotional project but which erases all frontiers declared arbitrarily by the obscuring intellect. This education of our sentiment and intuition does not orient toward some fantastic future age but focuses upon the immediate future. Yet rather than the imminence of the Final Day, this science postulates the mysterious omnipresence of Divine Perfection, its interior riches gradually displayed like the opening of a luminous fan. Essence always displays an infinite creation.

This science that continuously dissipates darkness is not mere academic knowledge or mere ecstatic experience. It is revolutionary action, outside universities, chapels, and catacombs—an action without the slightest trace of animosity, an action that resounds only with reverberations of Love, with affirmations of Truth. The illuminated ones who dedicate themselves to humanity, who engage themselves in Sufism, are not seeking refuge in scientific hypothesis or metaphysical speculation. Sufi science deals directly with ontological reality. The mature sages of every tradition who experience the eternal return of the One to the One, which is universal Sufism, repudiate all forms of skepticism and nihilism, as well as their basis, which is called, in every culture, the normal perspective of life.

This truly fecund and total education provided by Sufi science is not mere traditionalism. It does not indulge in regression to the past. To repeat blindly is the same as negating. On the other hand, Sufism is not eclecticism, with its complacent attitude that ignores the process of historical transmission, including its objective loyalties and disciplines. This science remains the center of convergence for all wisdom traditions, which are not interchangeable but are unique expressions of Divine Energy. From the panoramic perspective of this visionary science, one can appreciate the revelation of the same Logos through the living facets of human history and the living structures of nature. This alchemical science distills the wisdom and beauty of all Prophets.

Universality of vision does not weaken personal commitment to specific religious traditions, with their powerful sacraments and doctrines, their rich historical particularity. We are walking on a sharp edge here, but it is necessary to read each tradition carefully within the limitless context of the One. The devotion to a singular tradition is easier than the penetration into the essentiality of all traditions, but both practices are necessary. Each initiatory line must discover its irreplaceable position within the whole. No sage can avoid the responsibility and the adventure of this science of universal insight that always rejects half-truths. The ancient coat of arms of this noble science reads: *decipher, transmute, glorify*. These are complementary tasks.

Sufi science was the basis for ancient civilizations. It will also be the basis for future civilizations. This science, which is always contemporary, is the transformative manifestation of Divine Life—our sole root, the only root of the universe. The technical education of the modern academy commits the fatal error of hiding and suppressing our instinctive longing for rootedness in the Living One. Sufi education, on the contrary, manifests a redemptive character. Its principles and disciplines, which are more sensed than thought, witness to a mystical fervor and inspire one to meditate, not to make those summary judgments that are the dangerous coral reefs of the ordinary intelligence, gradually enclosing the soul rather than liberating it.

In all societies and all epochs, the conventional world, as well as its usual mode of education or its attempt to condition its participants, is always a scheme proposed from outside the soul, a strategy empty of prophetic vision. Prodigious in confusions, habitual social interaction is always full of antagonistic perspectives, menaced by invisible tendencies of violence and ignorance, which are perpetually disposed to explode into form. This dangerous conventional world crystallizes secret conflicts we have not resolved. Sufism, with awakened and intelligent attention, escapes this whirlwind of limited selves, this cult and tyranny of personality.

Sufi science practices sudden immersion in Reality, in the peaceful ocean of Essence. Following this baptismal experience, all attitudes and gestures of the Sufi lead toward the universal growth of love for God. Nothing and no one exists other than God. Nor is it necessary to dress this experience of Unity with the somber robes of a religiosity that suffocates spontaneity. Such religion impoverishes spiritual life, which should be a union constantly more intense and more joyous with the One God, Who is the most intimate of the intimate. Emphasizing the supposed distance or even abyss between Creator and creature, such religiosity is a rigid armor, a barren theory, a usurpation, an abdication.

The unique objective of Sufi science is to remain absorbed in God, generating a progressive illumination that ultimately pervades all humanity. This science differs radically from the ordinary play of perspectives, however sophisticated, which masks our pristine experience of Essence. Reaching perfect balance, with rectitude and with full consciousness of what it is doing, this science, like a compassionate surgeon, severs all our links with the discord between religions, cultures, and personalities. This is a genuine science, not a fantasy projecting some alternative world. Sufism investigates and expresses Truth, which is never influenced by our partial ideas and hidden weaknesses, either personal or cultural. Sufism is the vast moral, intellectual, and spiritual renewing of fallen humanity, a process initiated and sustained by the Logos, the Divine Word, the guidance that alone is worthy of our total confidence.

Blissfully free as it is from arbitrary forms, Sufism is not a field of anarchy, empty of spiritual disciplines, nor is it a hybrid composed of ideas in vogue, whether psychological or parapsychological. Sufism is not the project of a minority that wishes to impose its strange beliefs on the rest of humanity, nor is it utopian ideology or political revolution. It is necessary to make these distinctions rigorously. Sufism is, in truth, a great wave in human consciousness that creates history and that is the fruit of the aspirations of humanity in its totality. Sufism is not the exhibition of miraculous powers but the dignified inauguration of a magnificent reign of Truth, an experience of new light, free from the pride of those who fool themselves with their limited methods, limited sciences, and limited theologies. Experiencing the living Truth at the foundation of consciousness, the awakened person, the spiritual wanderer, takes shelter in this secret reign of Sufism, the most just kingdom that can be imagined.

The living power of Truth splinters all facades, dispersing into fragments all the sterile and empty projects—the prestidigitation of the controlling and separating self—that define the larger part of our assumed personal and social identity. This charismatic gift of Truth power flows freely in the ardent hearts of those who continue the transmission of Sufism. These are the spiritual guides who arise among every people on the earth. Having received the empowerment of Truth, they struggle and do battle without tiring and without resting, with great moral and intellectual vitality, with longing to penetrate to the nucleus of Truth and distribute its light and its power to all conscious beings. What is called Sufism is simply the inspiring perspective offered by the testimonies of these passionate persons of Truth—their poems, their conversations, their gestures, their brilliant silence. In open contrast with the anemia, inertia, and narrow-mindedness of the world and its institutions, these Sufis and their Orders exist in dynamic harmony with all consciousness, generating new light and opening the mystical path with friendship and clarity to seekers of Truth in each successive generation. While they already live in the inward condition of celestial existence, the virtuosity of Sufis upon this planet is astonishing. Since they remain solely under the influence of God, their sincerity and integrity is beyond doubt. After physical death, their holy tombs become spiritual gems in the earth. Their souls, which have disappeared into God, become seeds of light, planted in the sacred earth of the human heart. The Sufi submits all faculty of perception and control to the Only God. His or her spiritual passion inclines only toward disinterested action and love. Never pedantic or dogmatic, these perfected beings smile and remain silent. They are the revelatory texts of Sufism.

Sufi science remains free from any authoritarian or inquisitorial tone—free from any inflexibility—by simply bearing witness to Truth alone. In spite of the tyrannical reign of the domineering self, the surface history of humanity, Sufism flowers secretly within the radiant landscape of the heart. The mystical body of Sufism is not an organization ruled by juridical principles but a living organism that functions by renewing the conscious links of humanity with primordial Presence.

Careful investigation by means of Sufi science confirms that there is no arbitrariness whatsoever in the Divine Will. The universe could not be another way. To contemplate this science transfigures the nightmare of competitive relations between personalities and nations into the vision of a harmonious relation between finite parts and an infinite wholeness. This universal equilibrium is a network of lights in which all persons find their place as eternal souls. The existential consequence of this radical vision of the world is that human beings actually evolve into a different mode and dimension of living.

Sufi science brings into profound question our presuppositions about the limited self and the limited society, which are waves of psychic conflict rather than substantial orders of Being. Envisioning the ultimate consequences of truly human existence—in order to reveal with surprising clarity and directness the very One from Whom everything is born and to Whom everything is returning—this critical science erases the conventional notion of multiplicity as being a region, or even an existence, separate from the One. This compassionate science channels Divine Power into the Sufi heart, which can offer refuge to the disenfranchised—refuge where tears are wiped away and destructive passions calmed, where treason and skepticism vanish, where the irreducible Essence shines in total nakedness. This Sufi heart dares to manifest as an indissoluble current within the ocean of Divine Light.

Historical explications and systematically imposed intellectual structures are insufficient for understanding Sufi science, which transcends historical circumstances and passes beyond scientific, philosophical, and theological perspectives. Strictly speaking, Sufism is radically empirical. The experiments of this Sufi science are not laboratory tests, however, but tests of the heart, observations and investigations of the spirit. Sufism recognizes the value of critical sensibility and critical examination. It examines the suppositions that underlie mundane consciousness. It criticizes the pretensions of the ego.

Each Sufi heart—that is to say, each human heart that awakens—is invulnerable in its essential reality, its fullness without boundaries. Each Sufi heart is the secret and powerful ally of a vulnerable humanity. The Sufi heart possesses the sharp weapon of unifying knowledge, which cuts away emotional and intellectual passivity. Sufism does not exist as an objective structure but as the delirious enthusiasm of true hearts, free from anxiety and anguish—hearts that palpitate as the Divine Attributes of Beauty, Compassion, Majesty, Subtlety. Sufism is not a transitory or accidental occurrence. It is our universal matrix. It is our original land of birth. Possessor of total vision, Sufi science is not an instrument of knowledge but is Divine Knowing Itself.

Beneath the periodic variations in human history, sages discern this science of the One as the constant. But Sufism, the essential thread of gold connecting diverse civilizations, cannot be bought or sold, nor can it be produced in mass. It is the unique path of each soul to union with God, the path that never ceases to be surprising, because the One is infinite and never repeats Itself. In the open field of Sufism, no frontier separates spectators and actors. Each conscious being is a full participant in Divine Consciousness. There is nothing fixed or established in this drama of Sufism. There are simply the flowing, mysterious, and limitless possibilities of the soul—one with the Only Consciousness, one with the One. In the wonderful sanctuary of Sufism, there is no impatience, disdain, or irony but serene contemplation of the Original Face, a contemplation by which all discursive apparatus is reabsorbed into Oneness.

Although fluidly expressing itself through diverse cultural and religious forms, Sufi science is not constrained to express itself through any form. Whatever differences historical religions may conceptualize disappear in this impartial science, which opens the way to the mystical annihilation of egocentricity and which communicates through words of nearness and words of intimacy, through secret twilight language and wordless gestures. This rigorous science strips the person of vanities and pretensions, showing us the exit from a world without transcendence, a conventional world locked into itself. Sufism liberates the soul from the police state of controlling ego, from the atomic holocaust of biological death, from the Chinese Wall of apparent separation between humanity and Divinity. Sufism opens totally to God. No one and no thing exists for this science except God. We cannot live except in God and by God. Sufism regains the ontological innocence that is common to all beings, a blessed ferment full of the lightning flashes of mystic knowledge and the delirious downpour of ecstasy.

This science of incandescence manifests in a direct way that is both anterior and posterior to all intellectual operations yet completely satisfies the demands of the illumined intellect. Saving passion and sacred nourishment, this holy science is the opposite of modern sciences in their frenzy to know and control, a driving ambition that implies violence. This peaceful science is the reverse of all chimerical attempts, modern or ancient, to grasp, employ, or even enslave Reality. This science of the unconditioned manifests a totally distinct center and force of gravity than conditional sciences—a greater lightness, a greater purity of perception. The heart of this pure science is free from the deception and falsehood that popular imagination and even the scientific imagination claim to be real.

Developing and unfolding entirely within uncreated Light, Sufi science is comparable to the Banquet of Socrates and the Mystical Supper of Christ. Sufism attempts to live and breathe according to this extraordinary perspective—perfect companionship with Divine Wisdom and Divine Love. Even the most common daily events now assume profound meaning and carry with them great spiritual responsibility. Emancipated from the burden of rules, habits, customs, and norms, so painfully accumulated throughout human history, Sufi science reaches solid ground only in the One.

Sufism does not condemn the natural world. Its goal is not to disincarnate but rather to lift all theories and experiences of nature toward a coherent vision of the Divine Creation, always immersed, here and now, in the luminous ocean of the One. As revealed by ancient Sufis and their modern successors, the sphere of daily relations in all its detail is essentially sacred, a transparent expression of Divine Will, an eternal procession from the One within the One. Illumined by this incomparable science, free from personal evasions and cultural deceptions, the Sufi succeeds in transcending conventional religion and experiences true access to the path of Love and Wisdom, allowing his or her awareness to be guided by what is mysterious—rich in meaning, rich in spiritual rather than logical coherence. The disciplinary religion, as the disciplinary society or the disciplinary family, is an attempt to protect human beings against the violence and fear characteristic of the limited self. But the cardinal principle of Sufi science is non-violent victory over this arbitrary self: the lucidity to be able to face it and the valor to be able to dissolve it without residue into the One. Sufism is not some quixotic undertaking, some romantic trek to the edge of a precipice in vain search, but a precise and certain science by which to witness and verify all-embracing Oneness.

The fruitfulness of any society depends on the presence of Sufi science at its nucleus. Both male and female sages, transmitters of this science of contemplation and absorption, collaborate valiantly in defense of the various mystical schools called Sufism. They live to protect and perpetuate this point of focus, which remains totally open to transcendence, this clear spring of Divine Teaching, which wells up in the center of every human community.

The presupposition upon which the superficial norms of every society are established is the obsession with ego. This divisive ego is the fatal seduction: the enemy of true personhood and the exile from our essential homeland. The strange domination of this mannequin ego makes necessary the practice of contrition and penitence. The apparent or surface aspect of the human being, its rebellious egocentricity, is identified falsely with the real human condition. As a result, the abiding of mystics in the egoless essence of consciousness comes to be considered a transcending of the human condition. Not so. The unimaginable riches of selfless consciousness, the royal treasures of Sufi science, are flesh and blood of humanity. This is the true human condition. Each person contains in embryonic form the process of awakening to the essence of humanity, the process called Sufism. The obsessive self constructs an illusory wall between humanity and Divinity. The only duty of Sufism is to vault this wall. This leap is not a return to the past or a projection into the future. The Sufi does not even take refuge in the present, conventionally experienced as an isolated moment. When we succeed in untying the complicated knots of limited self, the earth moves beneath our feet. We are the Unique. We are perpetual Light, which diffuses through the mists of apparent limitation and dissipates them. We are now consciously marked by the seal of the Eternal. Saturated with intense joy, we overcome at last the false order of a world allegedly limited by death. We are the Source of Delight.

Nevertheless, the habitually constructed ego—insisting, with great tenacity, *I am, I am* and prolonging itself throughout history—is not abnormality, weakness, or treason, but simply the organic function of a nervous system. Over millions of years of battle for survival, the powerful structures of the limited self have become tainted with jealousy and suspicion. Sympathy and goodness are expressions of the eternal soul, which has opened a sacred way through this human nervous system—through limited selves and limited societies. The soul constitutes our real human condition, our true nature—vast and open, superior to all forms of tyranny, superior to the cultural and religious reductionism and imperialism called history. The soul experiences the complete concentration of multiplicity within the One. The soul is the mystic lover. The soul is the mystic path. The soul is our guide. The soul is our life. We are the single soul without limits.

The melancholy stoicism of the surface strata of all societies is the expression of the limited self—its superficial, automatic, mechanical interpretation of the nature of human life and the universe as a whole. But this negative view is a mere intellectual declaration or a mere emotional reaction of the limited self, absolutely devoid of Truth. Simply by posing with clarity and sharp sensibility this critical problem of the limited self and by demonstrating its mystical resolution, Sufi science resolves the conflicts within the human person and within collective social structures. This science liberates humanity from the exasperating and grotesque affirmation of its narrow personality, its abstract identities of race, culture, and religion, which are the forces manipulated by the charlatans and tyrants of all epochs. All conduct that is anarchistic and without compassion is rooted in fanatical adherence to the limited self—this obstinate desiring to be distinct or superior, this obsessive wish to occupy a post of great authority in the universe.

The limited self, both individual and collective, manifests as a constellation of ambivalent notions marked by belligerence, aggressiveness, territoriality. This apparent self, with its characteristic ability to sustain the appearance of solidity, can be masked in sentimental effusion or false empathy, simply waiting for the propitious moment to reveal its chronic violence, which is the doctrinaire and militant imposition of its own forms. In the face of this catastrophic reign of the limited self, Sufi science works patiently and silently to reestablish our conscious links with the eternal soul. Sufism distances daily awareness from the chaos of the aggressive ego, turning it toward essential consciousness, the ever-flowing spring of Living Water. Sufism returns to the Unity that exists here and now.

The divisive self is the sorcerer's apprentice. Its eccentric career produces war and desperation. Sufi science, with great critical penetration, distinguishes rigorously between true and false, limitless and limiting, pure consciousness and ego. This science is the soul's response to egocentricity, to this dangerous knot of contradictions—a response that does not simply criticize ego but transforms it by awakening from complacency, which is the ego's principle hiding place. Obedience to the limited self is idolatry, the temptation permitted by God to test and refine us. Idolatry is the crisis of the human species throughout its pilgrimage to the One. Sufism is the universal response of humanity to this crisis. Sufism is not an esoteric diversion, belonging exclusively to mystical practitioners in rare contemplative Orders. This common quest, this common breath, seeks and attains liberation and mystic union for humankind.

Consciously or unconsciously, the question all human beings eventually ask is this: *Shall we exist truly, or cease to exist?* It is not possible to diminish the intensity of this question, which manifests in the human being from its very essence as a living seed that must sprout. But the quest for true existence is, in a certain sense, indecipherable. There simply is no intellectual solution. Reality is not one of the puzzles designed by the limited self. It is possible to be totally astonished by the One but not to explicate the One. It is possible to be the One but not to enclose the One. Remaining suspended even for a moment in the contemplation of the One by the One, the limited self is dissipated like clouds by a tremendous wind.

The problem of life and death disappears. The chronic human schizophrenia, the dissociation between conditioned psyche and unconditioned soul, is blissfully healed. This reunification of all our fragmented perspectives is not reducible to logical reasoning. This elimination of the limited self does not imply the vanishing of the person in all his or her integrity and rectitude, entrusted with all possible responsibilities. After this illuminating integration of the temporal psyche with the timeless soul, which is simply a facet of the infinite diamond of the One, the entire network of social and personal forms, systems of rules and restrictions, chains of cause and effect—all become transparent to the One but nevertheless do not cease to be valid for daily life. In contrast with the psyche, the eternal soul, free from self, embraces the entire phenomenal universe as a message from the Divine Beloved, as a totality that emanates from the One while remaining within the One. The lines of separation between diverse zones of being are now seen to be illusory.

In his or her pilgrimage toward the kernel of consciousness, the Sufi never ceases to implore Divine Grace, without asking for exceptional favors from God and without seeking miraculous powers. The Sufi blends with God by abandoning the self to the Divine Embrace. Thus the Sufi actually becomes Divine Wisdom and Divine Love. The soul submits to God without hesitation or reservation. The soul loses itself in God by whirling around its essential axis. In the revolution of awareness that accompanies this conscious reunion with God and that extends beyond the frontiers of dream and waking, all pride, with its characteristic attitude of negating others, disappears within a single closing and opening of the eyes. In an instant that englobes every instant, the scenes of the universal drama accelerate and then evaporate. The great wind of the Holy Spirit sighs. The soul encounters no resistance, neither objective nor subjective. There is no more apparent congealing of Divine Light. The ordinary world is transfigured into a world that is in fact no world, beyond all limits of expectation, not passing through the filter of intellect. Only God can meet God. Only God can unite with God.

In the fullness of this meeting, this play of pure transparency, Divinity flows like blood through the veins of the exalted human body. Through the lips of His human beloved, God sings the mystical hymns of Sufism, radical and disconcerting poetry found in the profound strata of all cultures on the planet. The One God never ceases to make His Divine Beauty completely present through human consciousness, which is essentially the only consciousness, essentially His Divine Consciousness. Touched by Sufism, words suddenly cease to be emotional or conceptual and become a sacred river of Divine Energy, the water of life that sustains humankind and the entire cosmos. The language of Sufi science is a force of metamorphosis, similar to alchemical power, that transforms the very material of the universe into spirit. But alchemy is pursued clandestinely and narrowly, while Sufi science opens itself to all humanity.

The physical, emotional, and intellectual transformation experienced by the members of mystical Orders in the laboratory of Sufism is an incontrovertible empirical fact. It is not possible to disassociate this personal change of the mystic from the demand and the possibility of changing society. Sufism always initiates a dialogue with the whole society that requires a response from both sides. Through this potent dialogue, the limited self and the limited social and religious forms are already in the process of being overcome. Sufi science does not accept any linear conception of history. The point of culmination is always here and now.

May we sustain these universal postulates of Sufi science throughout our daily lives. May the sweet Mercy of the All-Merciful One be with the humble follower of the ancient doctrine who has written this essay and with the brothers and sisters of his illustrious confraternity, the contemplative and teaching Order of Nureddin Jerrahi, great mystic of seventeenth-century Istanbul, focus of unanimous admiration, his renown as totally illumined guide never ceasing to grow, even in the modern world.

May the peace of the unique Essence descend upon all Prophets, all human beings, all creation, for all is constituted solely by the One.

PERFECT HUMANITY Our presently emerging world civilization is perceived by Islam as the Muhammadan Age and by other noble revealed traditions in their own authentic terms. Our attention should not be placed on differences in formulation but on the gradually intensifying global experience of human unity, which mirrors the Unity of Reality. This awakening is predicted and awaited by all sacred traditions. Islam regards Muhammad the Messenger, arm in arm with Jesus the Messiah, as the inaugurators of this blessed age of spiritual maturity. Another meaning of *muhammad* is *insan kamil,* or perfect humanity, which is not primarily historical but is the timeless principle of Divine Self-Revelation. In the present essay, both these senses of the term *muhammad* are active and intimately fused. Our perfect humanity, prefigured by the beloved Jesus and unveiled and propagated by Muhammad the Messenger, is the Burning Bush, which we are each now approaching, as the noble Moses once approached living flames in the valley of revelation. This drama of encountering our own perfect humanity is now occurring not simply at special places of worship or in secret inwardness but within the entire life of planetary civilization.

This essay was inspired by reading the scholarly study by Luce Lopez Baralt, *San Juan de la Cruz y el Islam.* Once again, this effusion could not have been composed or even imagined by its author in English. It reflects the radiant inheritance of the Castilian language, perfumed by the Islamic spiritual culture of Spain. This Hispano-Arabic inheritance is displayed throughout the writings of Saint John of the Cross.

15
Perfect Humanity
The New Burning Bush

Prophet Muhammad, most beautiful recourse for the entire universe—may indescribable peace be upon him and upon his mystical followers throughout time and eternity—realized and propagated in the Arabian peninsula, during the seventh century of the Common Era, a union of various important initiatory lines. He propelled this living realization throughout subsequent human history, as the heart propels the bloodstream throughout the body.

Judaism and Christianity, fragrant with the gnostic atmosphere of the oriental desert, were richly blended within this prophetic heart, so wise and so loving, this great reservoir, storehouse and treasury of mystical wisdom. The Islamic vision of *insan kamil,* perfect humanity, blossomed spontaneously—a spring of Divine Grace eternally flowing within this brilliant prophetic soul, this crescent moon whose growth nothing and no one can impede.

The dome and minaret of Islam are the living proof of human completion. There now stream across this broad earth the four rivers of Paradise, revealed by the Glorious Quran: pure water of clarity, sweet milk of knowledge, intoxicating wine of ecstasy, incandescent and astonishing honey of the union between humanity and Divinity, between lover and Beloved. To experience and enjoy these four rivers here and now is the real meaning of our human birth—its perfection, its consummation. The fourteen centuries of this Islamic vision are like a dolphin of stars constituted by gnostic saints, or masters of Divine Wisdom, as well as by humble lovers of Allah, Who is the only Reality. Interpreting the meaning of this mystic dream, we could say that the luminous dolphin is humanity leaping from the roaring oceans of space and time into eternity.

This mysterious dolphin attracts, like a gigantic magnet, all eighteen thousand universes and their countless conscious beings, leading them to the Day without Evening, the Divine Luminous Wisdom. This Islamic vision of the culmination of Being and the completion of humanity comes only through Divine Mercy. Bearing in mind that we are not engaged in human speculation, there opens before the eyes of the heart the Muhammadan Age. Let us consciously readapt ourselves, investigating and penetrating this new frame of reference, this greenhouse for the new humanity.

Linking with all other living traditions on the planet and universalizing itself radically, this Islamic vision of perfect humanity will become the radiant pearl of the global civilization of our immediate future, a civilization in which we will consider the lives of other beings more precious than our own. With the joyous ease of Paradise, this universal Islam abides in the very cells of the human body, transforming secretly with its pure affirmation and expansion the molecular structure of everything that exists, rejecting and freeing itself from everything negative, mean, or base. This is the age of radical renewal, not mere amending but total transformation. This spiritual epoch unveils a vista unimaginable to the prelates and doctors of the secular church, which is modern science and technology, economics and politics.

O culmination of revelation! O Prophet Muhammad of the present—Summit and Seal of Prophecy, perfect human being who now takes up his sacred residence in and through all humanity, giving the final touch to its long spiritual evolution!

The crucial search for the mystical husband by the mystical

bride has now culminated in the power of the Muhammadan Gaze. All that is bestial and corrupt about humankind, all duplicity and egocentricity, no longer has any reason to exist. In the perfect humanity, which is now the active secret of our being, there exists not even the slightest harmful thought or intention toward any conscious creature because creation and its crown, the true human being, are revealed as a single concatenation of Divine Light. This is the instant when the curtain is being raised, when the cock crows at dawn.

In this Muhammadan Age of the perfection of humanity, the habitual scenario of conditional existence changes radically. All the way from each atom to each person to entire universes, Divine Reality is now revealed as intimately interrelated with human reality. This is not an article of faith but direct experience. We must passionately readjust our antiquated cultural and spiritual frames of reference. This sublime earth, along with countless other inhabited worlds, presents the spectacle of a great airport where each day millions of souls arrive and leave, through birth and through death. Their compatriots are weeping for them with luminous gnostic tears—tears of the recognition of light by light. We must come forth from our circuit of habitual and conventional meanings and arrive at the essential point of our being, delighting in the present completion of humanity and creating bonds of love with all human beings in this new humanity. This is not a mere change of costume or theatrical role but a permanent, new constellation.

This unexpected interpretation of the sacred scripture of our humanity, our mysterious elevation to Divine Perfection even during earthly existence, comes from the unique Light of Guidance, which has flowed through all noble Prophets. Access to our true nature is opened by continual remembrance of living Light—infinitely conscious, beyond analysis and commentary, even beyond impression. The mind is not capable of seeing this new human reality except through the eyes of the heart. We are not mere spectators but profound communicants of perfect humanity.

In this central Islamic mystery, the completion of humanity, there is no assignment of fixed roles or meanings. There is no distinction of race, gender, nationality, religion, class, or caste in this family of the One Consciousness, this mystical Order that possesses countless branches. God is no longer revealed as a subtle ray shining through the person. It is now the whole person who is revealed as exalted in God. This is the transformation of sand into pearl, carbon into diamond.

The final meaning of Islam, universal peace, cannot be superimposed by personalities or by religions but is introduced directly through perfect humanity, which is now palpable within each heart. In this mystical courtship, we follow within our own consciousness the footsteps of the perfect human being, repeating with each breath, "Your presence amazes me, annihilates me, converts me into diamond." Outside the transcendent Attributes of God nothing really exists. These Attributes, or essential Divine Energies, exercise all the activity of countless universes. These Divine Attributes are not some other reality but are the true nature of humanity, or Muhammad, in its state of culmination, called universal Islam.

O Allah, in Your sublime Court, please crown us as the Sultan of Your Love, as perfect humanity!

The Divine Will sealed this perfection of what is human in the common planetary year 632. The long apprenticeship of humanity then reached its completion. The potent spiritual antitoxins against the contaminants of the warped ego and against the contagions of the deceptive society are already active and effective. Free from deprecation and disdain, we may now wear the white robe of the mystical resurrection upon this earth. Seated on our knees, with open hands resting tranquilly on our thighs, we now orient toward the secret Mecca of the essential point, which is perfect humanity. We are irresistibly filled with the powerful, illumined insight that only God can merge with God, only God can be God. We now emulate the perfect and imperishable humanity of the Prophet

Muhammad, his actions and his loving words, enjoying together the exquisite atmosphere of Paradise during this earthly life. It is no longer necessary that we barely survive upon the crumbs and crusts of the conventional world.

O Wisdom and resplendent Primacy Who has finally dissipated all obscurity!

This perfected human reality is not solitary but communal—a new grammatical syntax, a new comradeship. All of us are mystical verses of the Psalms. Perfect humanity is the Song of Songs. We breathe the atmosphere of ceaseless rejoicing, which is the Resurrection of Love. We become pioneers in the crafting of spiritual gold, which is the body of golden light, the body of the resurrection. Filled with mature enthusiasm, we exchange brilliant gifts of love. Donning new vestments—new spiritual states—with each heartbeat, we live in the mystical conviviality of our new humanity, with no more veils of individuality, substantiality, or separation. We actually interchange personal beings with one another, allowing our highest aspirations of concordance and communion to be realized in a totally unsuspected manner. The beautiful interior form of the human being is a unique configuration of the Divine Attributes. Within this formless form abides the consciousness without shadow, which is Divine Essence, the heart of perfect humanity.

We are sprouting seeds of this new humanity—anointed ones, named ones, dedicated ones, the arabized friends of Muhammadan Light. Our mode of living in this confluence of humanity and Divinity is entirely distinct from the usual mode of existing in a world that is narrow, solidified, and stipulated, in a bittersweet and capricious life that is not really worth the pain of living. We now breathe and dance without care, barefoot and delirious, in the presence within ourselves of the new Burning Bush, the Muhammadan Light, the incomparable and immense treasure of the true human heart. Everything that has ever existed or will ever be is active within this living treasure of the new humanity.

O loving confidants, O mystical colleagues! Let us freely and generously distribute indescribable spiritual riches during this age of fulfillment!

Free from subtle rancor, the poisonous serpent of warped ego, the newly unveiled human heart can adjust gracefully to all forms of praise, prayer, communion, and contemplation revealed throughout history. We are the all-embracing mystics who belong solely to the Reality that has no comparison. We exist without the slightest shadow of superiority, without the slightest sense of condemnation of the creation of Allah or any of its precious creatures. Everything has been placed within the spiritual reach of perfect humanity, which is now our own true being. Our sole duty is to gather the honey of Divine Life in companionship with our compatriots and collaborators in the universal Islamic vision—passionate persons free from hermetic, ascetic, or elitist tendencies, persons who express themselves with the unequivocal clarity of the radical equation *human equals Divine*. With this honey, we feed the universe.

According to a beloved Sufi master of this century, a living library of mystic knowledge, Bawa Muhaiyaddeen, to whom we offer our loving gratitude, Divinity proclaims, "I am the secret of humanity and humanity is My Secret. Humankind is My Wealth and I am its wealth. I am humanity and humanity is I." There remains not the slightest doubt. There is no place left for cultural or religious provincialism. There is no middle ground between extremes, because there is no duality.

O Christ the Messiah! Your gentleness, which offered the other cheek, was the dawn of perfect humanity. The royal vestments of your mystical body, which is our own spiritual body, glow with the morning light of universal empathy. O most beloved Muhammad! The new heaven and new earth inaugurated by the tender Messiah are now established firmly upon the base of your great love, with the spiritual support of your beloved ones throughout fourteen centuries—who dance upon this sacred earth and within Paradise simultaneously, remaining fully immersed as well in the formless Garden of Essence.

This ultimate union of humanity with the Source of its being is confirmed by consensus, by the collective agreement of the sages on this planet who proclaim the universal Islamic vision—those who are versed, those who are adept, those who are in love. These intimate friends of Allah are daring, illuminated, realized. They move forward with intellectual honesty—free from any limited pretext, any complicity with the world—expressing themselves through personalities that are provisional and dispensable. The Attributes of Allah are replicated in the entire being of such persons, who have been made malleable by Divine Energy.

O Muhammad, the image of seven concentric interior castles used by our noble mother Teresa of Avila expresses your mystic way of the concentricity of the soul and God. You are the inheritor, admirer, and conscientious transformer of the mystical teachings and contemplative practices of ancient Christian monasticism. As Saint John of the Cross proclaims, "These grades or wine cellars of love are seven." Such are the seven levels of consciousness traversed by the interior science of Sufism. As confirmed by Muzaffer Ashqi al-Jerrahi, a sublime Sufi master of this century and a humble servant of the new humanity who vested thousands of persons with initiatory power, "Upon this green turban of the Dervish Order, all seven heavens are resting."

Your undeniable presence and subtle imprint, O Muhammad, entered the spiritual center of European civilization when Europe voluntarily explored Islam during the Middle Ages in order to counteract the powerful cultural pressure of Islam at her borders.

The voice of perfect humanity testifies: *apostasy from the Religion of Unity is impossible.* No conscious being can escape the all-powerful attraction of the basic simplicity that is Divine Unity—the sole Consciousness, the sole Reality. It is not possible that the drop of dew on the petal can resist the attraction of the sun. With surprising care and intimacy, infinite Light whispers to the drop that balances, trembling in ecstatic concentration, "Please do not hesitate! I am longing to receive you in the most sweet reunion." The ground of the universe is pure Love—supreme Beloved without intermediary, Guide for all souls, which by their original nature are absolutely uncorrupted. The conscious drop does not disappear but is translated into pure Light.

O Allah! O Infinite Beauty Who captivates the universe! Finally, You have placed the jeweled crown of Muhammadan Light upon Your most precious humanity, pouring indescribable blessings, the praiseworthy acts and thoughts of all the Prophets, into each human heart. Our entire being is soaked by the rain of perfection. All success and realization comes only from Allah, may He be ceaselessly praised, Who has inaugurated this Age of Muhammad at the end of evolutionary time. This Muhammadan reign, this kingdom of humanity, abides as the fullness of a temporality completely open to eternity. Profound creative and explorative liberty awakens. It is not an end but a beginning. This new humanity is displayed like an opulent royal robe, shining with multiple mirrors, reflecting secret congruencies between all ancient religious and esoteric languages. We become those who can speak every mystical idiom. We are entrusted with speaking the tongues of revelation.

Adorned by this robe of spiritual royalty, trimmed with jewels of various surprising faculties, the new humanity experiences a rich sense of generosity toward all creation. It is not merely a change of dress, but the rapturous fusion of human and Divine. The puzzle of everything that appears illogical or paradoxical disappears in the blaze of Muhammadan Light. For perfect humanity, there is no birth or death or rebirth. There is no longer the slightest sense of separation between reflection and Original.

Human sensibility has reached its long anticipated final integration. There now stands revealed what the Messiah Jesus called a *table in my kingdom*. The earthly garden of perfect humanity is refreshed and irrigated by Divine Rain, by the most delicate elaborations of the immutable principle of Unity. Proclaims Nuri of Baghdad, a classical Sufi of the Muhammadan Era from more than a thousand years ago, "Allah, may He be praised across the entire face of the earth, manifests a secret, inner garden. Anyone who smells its fragrance can no longer desire any other Paradise. This garden is the heart of the mystic, here and now." To enter the Garden of Essence and travel there with spirit entirely transported, during both earthly life and eternal life, constitutes the new nature, the new focal point of Muhammadan humanity. As prefigured by the Christian mystery, this true humanity is the marrow of all worlds, visible and invisible.

O heart of perfect humanity! Our own heart! Beneath the freshness of your arcades, within your tranquil, sun-filled patios, may we rejoice in the single heritage and brilliant diversity of all Divine Messengers. May we make daily prayers in the radiant shade beneath your smooth domes, within your vast cathedral space. Strengthening and calming the breath, may we intone beautiful Divine Names with each respiration in the sacred mosque of this new heart. May we live in constant expectation of the signs of new humanity, even in the most simple and intimate details of our lives. How else can it be in the embrace of perfection?

Surprisingly, even in this era of spiritual completion, we are still armed knights with courteous manners, protected by coats of mail, consecrated and committed to the common task of holy war against unfocused and superfluous passions, characterized by stubbornness, torpor, coarseness. We must resist these virulent passions because of their aggression against the coherent vision of our new humanity. So there still exist chivalric Orders of knighthood, Orders of hospitality, which bear the coat of arms of Divine Love. Their initiates are seated knee-to-knee with the most chivalrous Muhammad, not below him. They live face-to-face with the true human being, observing his aura of holy flames, repeating in ecstasy his epithets of praise, rubbing their faces in ecstatic devotion against brilliant golden sand in the Desert of Love. These perfected ones compose wonderful epistles to the entire world family, praying, "May the final breath of every person be one of the innumerable Divine Names. May the love of the lovers of humanity augment day by day. May Muhammadan Light burn to ashes all negative characteristics, purifying and awakening limited humanity into the limitless landscape of perfect humanity."

Allah, praised and glorified be He above all conceptions, does not admit of any comparison. Yet the secret heart of humanity, awakened and gradually unveiled through the instructions of all Prophets, is comparable to the Divine Heart. Humanity is the perfect mirror, which we simply must cleanse from tarnish and vapor. With our arid personalities flooded by means of spiritual aqueducts, our subtle interior system of irrigation, the sanctified humanity of the Blessed and Most Pure Virgin becomes our own humanity. Her miraculous pregnancy by the Spirit becomes the fetal presence of perfect humanity within the womb of our heart. The entire history of revelation refers to the mysterious birth-giving of this totally transvalued humanity that enjoys limitless spiritual reach. It is the promise of the beloved Jesus Christ and the realization of Muhammad, Mercy to all Worlds, crest and culmination of humanity. The last trace of separation between what is truly human and what is Divine has been erased by the gnostic saints of Islam, each one a brilliant star upon the earth, guiding us to the mystic birth in the cave of the heart. Each one of these merciful sages is a magus, a wise king from the East. Each is a virgin mother, each a messiah of the new humanity.

The supreme secret is no longer inaccessible. There are mature persons who clearly perceive the new heart. Soaked with continuous rains of Divine Mercy, this paradise of the heart has no circumference, no frontier. The veil before the holiest sanctuary is torn away. The temple constructed by human hands and minds has disappeared and is replaced simply by perfect humanity. The immemorial prophetic teaching, recognizably Sufi, has reached its final evolutionary goal in this Age of Muhammad. The human being can now live in perpetual, conscious union with its supreme

Source. This infinite conflagration of Divine Love consumes human reality and transforms it into diamond with the constant call, "Be the perfect human being!" This mystic completion was prefigured by the adorable Messiah Jesus, who cried, "Be perfect, even as your Father in heaven is perfect!"

Sings Saint John of the Cross, therapeutic poet of this great universal transmutation, poet of illumination and abandon, "O sweet fire, O sweet flame that ignites hearts and converts them into pure love, O living flames of love, O fire of Divine Love, which transforms us into God, which deifies us." The mystic night of Saint John and his brother and sister Sufi lovers has opened a way for hearts filled with Divine Love, an access to direct experience of Divine Light—intuited as flames, lamps, auras, lightning flashes, spiritual dreams. As Allah revealed to Muhammad, His consummating Messenger, "I am a Hidden Treasure Who desires to be known." This Divine Self-knowledge manifests as a night—dark only because of its absolute brilliance—in which the nakedness of the soul is achieved, the unveiling of perfect humanity, freed from the limited perception, appetite, avidity, and indulgence that we call the world. The dark night of the soul is not depression but sublime exultation.

By means of alchemical alternation between states of spiritual constriction—longing, weeping, emptying—and states of spiritual openness—ecstasy, delight, expansion—the mode of ascent, which Saint John calls the *secret ladder,* is displayed before the eyes of the heart. Three successive dimensions are revealed: early night, midnight, and dawn. This is the parallel in the teachings of Saint John to the threefold Nocturnal Journey of Muhammad, who was translated at night to the holy city of Jerusalem, who then ascended by means of a ladder of light through the celestial planes to the bright midnight of Divine Essence, and who finally descended again upon the earth, bringing the Good Pleasure of Allah and the prayers of Islam, which are a mystical ascension for all humanity, the dawn of perfect humanity. The history of this Nocturnal Journey, this crest of the wave of prophecy, was well known and carefully studied by scholars and mystics in the European civilization of the Middle Ages under the title *Ladder of Muhammad.*

Each person who awakens into this era of Muhammadan Light—his or her very breath become pure Divine Peace—fulfills precisely this same responsibility of mystical traveling, both ascending and descending. The Sufi Orders, through their initiatory chains of succession carried on by realized shaykhs, possess authentic knowledge of this secret ladder. They continue to invite humanity to unveil the soul, to discover what perfect humanity really is. One of these peacemakers, one of these contemporary Sufis, sings to the beloved Muhammad, "Your enamored gaze is my sole abode, my sustenance, and my breath." It would be an error to consider ourselves as apart from Muhammad, as apart from perfect humanity.

Conscious cells of the new humanity, gold miners in a Golden Age, Sufis attain the experience of planetary citizenship, a theophany across the face of the earth. All opacity, including ideological and political frontiers, disappears within the perfect transparency of Muhammadan Light. New humanity now constitutes an integral and intimate community, founded in spiritual affinity, breathing through interchange among all the noble wisdom traditions of the planet. Such universal conviviality is as natural and basic for the new human being as air and water. This new sensibility is free from emotionalism or superstition. What now presents itself spontaneously is a sacred tradition that is all-embracing yet multipolar or plurivalent, a tradition liberated from legalistic persuasion and catechetical indoctrination.

What is occurring is not a graft or transplant from the conventional world. This is nothing less than the Kingdom of Heaven upon earth—the cornucopia of true abundance, love without self-interest, freedom from the slightest trace of possessive ego. It is both the promise of Jesus, the sacrament of his presence, and the gift of Allah through Muhammad and millions of Muslim men and women, immersed in family and social responsibilities yet awakened to the most intense degrees of unitive wisdom.

Perfect humanity need not retire to any ascetic and isolated hermitage, to any cell of celibacy. Concrete daily life now becomes love for Divine Love's sake alone, passionate lyrical breath, high point of lived poetry, profusion of enlightening images perceived with spiritual wakefulness, whether during dream or in the waking state. Rejoicing moment by moment in the fragrant company and beneath the delightful gaze of the Beloved, experiencing the delicacy of divinely human love even in the midst of tribulation, the lover of perfect humanity never laments. To live is to decipher this fluid allegory of every day, pregnant with ultimate meaning. To live is to receive the superessential bread of every day for which the Christian lovers pray in the *Our Father.*

The Trumpet who announced the dawn of perfect humanity, Christ the Messiah, provided the purifying experience that made us capable of attaining, here and now, within this human vessel of awareness, the transforming union with Omniconsciousness. O irresistible and captivating Love! O subtle rejoicing of the contemplative life, which now ceaselessly unfolds in the secret heart of perfect humanity! O drunkenness and tireless longing of the entire universe, following and kissing the steps of the perfect human being! May we drink the wine of Love offered by the powerful right hand of Muhammad as an unbreakable covenant, enjoying the divine company of perfect humanity. May we be constantly lost in ecstasy, in the mystical state of unicity. This is the very nature of our new humanity. As a contemporary Sufi sings, afflicted by Love, blazing with flames of Love that could melt an entire glacier, "Simply the mention of your majestic name, O Muhammad, leaves me overwhelmed, silenced, amazed." As the ancient Sufi Niyazi sang, "Nothing can hide the Face of Truth." This face is perfect humanity.

O illuminated philosophy from oriental sources with Islamic components, diffused by various channels throughout European civilization! O luminous wisdom that displays itself in all cultures through the long history of humanity! O exuberant dervishes of every religion, secret coreligionists, prisoners of Supreme Love who pass the night in vigil with hearts palpitating! O heaven upon earth! O completion of humanity!

After the political failure of the Crusades, all Europe subtly acquired the intellectual, artistic, and spiritual heritage of Islam—through the mobility of merchants and troubadours and through the labors of various secret intellectual intermediaries and famous promulgators of Arabic culture such as King Alfonso X of Spain. The consummate European mystic Meister Eckhart regularly quoted from four Islamic philosophers. Protected by a cultural amnesty, the new Arabic knowledge manifested without difficulty through the entire Middle Ages by means of systematic translations from Arabic and through the hidden inclinations and affections of Sufis—Christian, Jewish, and Muslim. More than a fertile influence, this process of voluntary Islamization constituted a complete transmission. Christianity, Judaism, and Islam were interlaced without partitions—with intellectual honesty and religious integrity—drawing their esoteric resources from a common treasury. It was the arrival of seventy camels laden with silver and gold, beneath a mysterious rain of diamonds and pearls, rubies and emeralds.

In the center of this cultural and spiritual interpenetration remains the impeccability of the Prophet Muhammad and his way of imageless meditation—the way that maintains direct affiliation with the mystical teaching and method of the Orthodox Church of the East, the Christian mothers and fathers of the desert and their experience of the uncreated Light. This transmission, hidden or perhaps even lost in the sensibility of the Western Church, is our cardinal vocation—deification of the whole being, not only after death but also during this earthly life.

Men and women completely illuminated by Islam sing about this mystery of deification through imagery of the lover attracted with great intensity and intoxicated by the perfumed black curls of the beautiful One. The lover transcends these brilliant curls of creation and is lost in timeless Divine Beauty, gazing at the delicate and peaceful face of the supreme Beloved. This disconcerting mystical attitude, which identifies passionate human love with Divine Love, Sufis call *smashing the bottle of shame and modesty*. The true lover would not hesitate to sacrifice everything for this experience, crying out with abandon in climactic communion, "I am the Beloved." This is the declaration and unveiling of perfect humanity. Glorifying Allah with this divinely human love, the Sufi, at perfect ease beneath the brilliant green canopy of the Prophet, lives with candor, free from all recognizable parameters or paradigms. These unclassifiable lovers leave circumscribed society disconcerted, yet they become a source of delight to true hearts everywhere.

The teaching of supreme union—which the mystical theology of the Christian East attempted to articulate, experiencing both intellectual and political difficulties—is expressed by most Sufis through silence, like the beautiful lilies of the field in the parable of the Master. Other Sufis express this radical teaching in secret code, through their passionate poetic experiments. In this final stage of union, or deification, all language fails. This superconscious state is beyond bridal or bacchic symbolism. With awareness entirely awakened and miraculously expanded in the fusion of abandon and equilibrium, Sufis manifest the dwelling in Paradise upon earth, which is true humanity. This demonstration of Paradise is rooted in the holy science that triumphs over death and that lifts us from speculation to verification, from devoted faith to gnostic vision. The cardinal principle of Sufi science is that awakened humanity is Divinity. As was explained by the great scientific Sufi Ibn Arabi, whose teaching descended mystically from Muhammad, "God thinks through me, because I am God thinking." We are absolved from the burden of any limited identity.

Deepening every moment in the delight of omniconscious union, this unitive identity nourished by the delicacies of

Paradise—this Muhammadan humanity—is a single soul with multiple forms. The one mystical body of humankind responds to and consciously expresses the infinite longing of God to taste, feel, and be His Own Reality. Such is the unifying knowledge of the Seal of Prophecy, the beloved Muhammad, may Allah bless him and give him peace—Pearl of the Universe, Green Hummingbird who Sips from the Rose of Wisdom, Secret Heart of Humanity. This universal human heart Saint John calls "a Paradise of Divine rejoicing."

O mystic sultans and sultanas who elect the throne of the heart, preferring it to the throne of empire! O predestined ones! O consecrated ones! O spirits who are ecstatic even when undergoing privation and tribulation! O flowing water transforming the wasteland of despair! O flowering orchards, vineyards, delightful gardens! The Paradise meditations that spring from the Islamic vision, rooted in the Holy Quran, are effusions of those on the path of sanctification and deification. Here is intellection that is vision, union, and fusion with uncreated Light.

The operative key in this planetary symbiosis of cultures and traditions, their convergence as a single finality, is the active intellect—identifiable as the Divine Light without dilution that inevitably manifests as the Age of Muhammad through a humanity now fully evolved. It is not necessary to approach this new Burning Bush with precaution or timidity. It is our own most intimate light.

In this absolutely new frame of reference, which is humanity freed from its dictatorial and divisive ego, there can no longer exist fanaticism, intolerance, or inequity. It is submission in love without subjection to law. Islamic jurisprudence and all other sacred law culminates in the Sufi mode of behavior, which is simply to embrace love, to catch flame in love, to pour forth love to all creation. Error is nothing more than a lack of love and is for this reason always condemned to failure. Let us take the text of our existence in its mystical sense, translating human reality into

Divinity, mute with amazement in the face of our own true nature, in the face of perfect humanity. Our history, both personal and planetary, is a spiritual epic: Perceval in search of the Holy Grail. No authoritarian or dogmatic explanation is appropriate. There can only be a visionary interpretation of this drama, this quest that eludes every facile description. This tumultuous ocean of divinely human love is Light within Light within Essence.

O aspirants to the new human reality, keep your critical spirit alert, your discernment and perceptiveness so clear that nothing and no one can cloud your unitary vision. Struggle arduously in full battle against malignant spirits, which would block the road of the new humanity. Satan, the joker, impostor, parasite, and cynical assassin, is empty of true substance and can only snarl and scream with his deceptive conceptuality, his habitual obsession, his cancerous inanity. His illness is contagious, his evil infectious. But be absolutely certain, O lovers of Truth, that authentic spiritual combat against the adversary of humanity, against the demon in all his corrosive forms, is always victorious, thanks to the intervention of realized beings, who are the force of the new humanity.

The secret meaning of this novel of chivalry that is our own life is freedom from self, dethronement of ego. As Saint John of the Cross explains about his own spiritual emblem, which he calls the *colorless bird,* "Do not take initiative in anything." This principle of total renunciation extends even to mystic knowledge. Sings Saint John, the solitary bird of omniconscious unity, "Only know God, without knowing how." To feel excessive attachment to the forms of society, including religious forms, is to convert them into an obstacle or a veil between the aspirant and perfect humanity.

At the peak of his ascent of Mount Carmel, Saint John proclaims, "To here, there is no road." The scholar Luce Lopez Baralt explains in her admirable study of Islamic roots in the poetry of Saint John of the Cross, "The saint carries out a circular and nonexistent journey from God to God." This is the Sufi journey within God and beyond God through mystical annihilation, through the ecstasy that obliterates. The only heresy within the new humanity is to attach oneself to anyone or anything that is not God. As Saint John affirms, "One single thing is necessary, which is to know how truly to negate and annihilate oneself." There is no question here of nihilism. On the contrary, as Shaykh Muzaffer of Istanbul confirms several centuries following Saint John, "It is an absolute certainty that those who are annihilated in Allah will exist forever."

This arduous and ardent ascent of the spiritual mountain is not possible by means of didactic or cerebral methods. Nor is it possible to detail or systematize the path with any exactitude. O lovers worthy of the most sublime title of perfect human being, observe the fleeting traces that manifest before your feet. This is the beautiful path of uncreated Light. The mountain is at the same time a fountain, a spontaneous spring—"fount that emanates and flows," as the delirious Spanish poet sings. The unveiled peak of this mountain is perfect humanity.

Another Sufi poet and essayist of the Iberian peninsula, Ibn Arabi, declares, "When my supreme Beloved appears, with what eye will I gaze upon Him? With His Own, not with mine, because no one can see Him except He." An almost insupportable ecstasy, a rupture of the logic of the world, a leap that surpasses religion, must occur to capacitate us to see God in the unitive manner. The iron plunged into fire turns vivid red.

The central point within the ceremonial circle of the dervishes is called the gallows of Mansur. The beloved Mansur al-Hallaj, chief of authentic lovers of the sole Reality, loved so much in the unitive way that his eyes never saw anything other than the Truth. In this state of blissful forgetfulness of himself, he cried with the voice of the perfect human being, "I am Truth," which in the Islamic context plainly signifies "I am God." Condemned to die for this apparent blasphemy, Mansur encountered his physical death upon the gallows of the limited society, through its limited commands. He is a paragon of mystic love for all those who would lose themselves ecstatically in Truth, dying to all confining conceptuality, even that of religion. Perfect humanity exists at the center of the circle.

The unitary experience of the friends of God throughout history abides at the root of planetary mysticism. This kaleidoscopic path—elusive and inspiring, accelerated and passionate—is both the physical and metaphysical contemplation of Divine Light, which opens us to unsuspected dimensions of our own humanity. As Saint John of the Cross, fraternal twin of Ibn Arabi, sings, "A most clear door opens in the manner of a lightning flash."

The heart is a spring at the center of a clearing within the uncharted forest of creation. Here, what is human, irradiated by Divine Love, transforms into what is Divine. There is nothing other than God, nothing other than perfect humanity—which is simply the conscious realization that God alone exists. Successors of Prophet Muhammad from across the entire globe have torn away the veils that covered Divinity, that apparently separated Divinity from the human being. "Lover transformed into Beloved," sings Saint John.

The advanced guard of perfect humanity are not the knowledgeable scholars of Islam but those who represent the living logic of the Islamic vision, in whatever tradition they may appear. Amazing indeed is the avalanche of mystery provoked by these incendiary beings. The mirrors of their hearts well polished to reflect the Muhammadan Light, these persons beautify and sanctify everything with a freshness that never fades, a love that endures forever. The unveiling of this brilliance of the perfect human being consumes the mundane layers of opacity that impede our total encounter with the sole Consciousness, Who can call Itself equally Allah or Humanity.

To the question "Will your soul enter Paradise after death?" Saint John Chrysostom, Christian Sufi and Bishop of Constantinople fifteen centuries ago, responded, "I am not concerned with the journey to Paradise. I have become Paradise." This is exactly what the other Saint John indicates when he speaks about "the transformation of the soul into God." The western mystic poet cries out in the delirium of union, "O abyss of delight." The suffocating sensibility of conventional religion cannot exist in this abyss of omniconscious unicity, where there is neither professorial chair nor any other organizational base for personality. To enter this luminous abyss without frontiers is truly to complete the pilgrimage to Mecca. To contemplate the brightness of this abyss is truly to understand the Glorious Quran, the Final Proof, the password to mystic union.

O members of the new humanity, let us consecrate ourselves without vacillation, with insatiable longing, to the unveiling of human perfection, whatever the apparently negative circumstances of history. Let us disregard the false claims of the conventional world. Let us overcome the spiritual exhaustion of our particular society and its institutions. Beings of light long only to blend into Divine Light. This longing is the true fast of Ramadan, a fast from all concepts that limit humanity and Divinity.

O Allah Most Near, may hearts occupy themselves solely with You. The perfect humanity hidden within the human form is the Supreme Self, diffusing Itself with shining love into the entire being of the lover. O Divine Beauty, nothing other than You manifests within or beyond creation. Divine Creativity is the One returning to Itself. This is the neoplatonic circle of emanation and return, enriched by the prophetic vision, which does not annul the value of creation but witnesses its spiritual transfiguration and elevation. This return is not a regression to original Unity but an advance to perfect humanity. There is no being other than the One Beyond Being—the Nonbeing, the Abyss, the Heart of Reality, which ultimately displays Itself as perfect humanity. O Allah, enlighten humankind to its own true nature with Your Light of Guidance—the light of universal intellect, the mystical certainty of perfect humanity.

O passionate Sufis, how intensely the correspondences augment in this luminous poem of your lives, this poem worthy of the sublime title *Perfect Humanity.* Maintain carefully within your hearts the lamp of Muhammadan Light, the inextinguishable love that illuminates the most secret caverns with a sun more brilliant than midday.

Welcome to the eternal Day, so gentle that the camellia remains damp with dew. We are feasting here and now in the company of the most beloved Muhammad. We are falling prostrate before his Divine Beauty, which is our own. The mystic night of unknowability has disappeared. Mystical intoxication has reached its end. The Divine Emanations, including all possible worlds, have returned to their Source. The fascinating enigma, plurivalent unicity, has been resolved. Royal peacocks display brilliant colors and mystic poets chant from the infinite Song of Songs, the Book of Humanity, inundating their veins with the ardor of audacious verses. This is the universal unveiling of perfect humanity. Creation manifests as a ripe pomegranate, its brilliant seeds realized human beings. The Creator, Love Itself, is now openly revealed as the fervent lover of perfect humanity, the history of prophecy as Divine Courtship. Love now proclaims the most panoramic criterion, in the light of which all possible systems of explication dissolve. Conventional structures are dismantled. The upside-down society is abolished.

The Holy Spirit overflows all channels, divulges all secrets. Spiritual treasures that transmit themselves without personal agents overflow like milk and honey. It is what Saint John of the Cross calls "the mystical supper that recreates and enamors." This is the ecstatic uproar of holy lovers, leaping beyond the fundamental laws of logic. This is the unsheathing of the sword, the parting of the curtain. As Prophet Muhammad proclaimed in the voice of perfect humanity, which is our own voice, "Allah is the eye with which we see and the tongue with which we speak." This is gold without dross. This is the Muhammadan flood tide.

Exalted spirits awakened from rapture, without fatigue and with rapid steps, go forward to the station Saint John called "to understand without understanding." Established beyond the historical books of revelation, freed from the obliqueness of cryptic and hermetic language, this direct experience could not be more full. It is human reality without alloy, liberated from dogmatism, flowing with goodness and delight, which are indescribable because without any possible comparison. This is the transition without end and without beginning, absolute elasticity and perfect ineffability. "O crystalline fountain," sings Saint John. "O gentle wine of Love. O Divine Fragrance."

O fragrance of amber in the court of the Adorable One. O colloquium of God with God. O mystical bed of lover and Beloved. O traces of Quran in the infinitely faceted mystic love poetry of Sufis within every culture. O irreducible mystery of Muhammadan Light. O ancient dynasty of Divine, Luminous Wisdom. O passionate beauty of perfect humanity. O uncontainable passion for communion. O heart of the lover in flames, in ruins, destroyed by ecstatic love. O tender names of the Prophet Muhammad. O Muhammad, Seal and Culmination of Prophecy. O amorous secret of the new humanity, the new heaven, and the new earth. O abiding of Paradise upon this planet. O perpetual dawn. O conscious state of early morning. O effervescence of the entire universe. O ardent lyric poems of love. O effusions of wise poets, which expand the heart. O fully evolved persons in the Age of Muhammad.

Most precious Allah, we can do nothing. We cannot survive even for a single instant without Your sweet Presence, expressed through the perfect human being, *insan kamil,* mounted upon the noble white steed of universal Islam.

Index

QURANIC REFERENCES